EXPLORING ROCKS, MINERALS, FOSSILS IN COLORADO

BY RICHARD M. PEARL

REVISED EDITION

SAGE BOOKS

THE **SWALLOW PRESS** INC.
CHICAGO

Revised Edition

Sage Books are published by
The Swallow Press Incorporated
1139 South Wabash Avenue
Chicago, Illinois 60605

ISBN 0-8040-0105-7
LIBRARY OF CONGRESS CARD CATALOG NUMBER 64-25339

Book was originally published under the title,
Colorado Rocks, Mineral, Fossils.

EXPLORING ROCKS, MINERALS, FOSSILS IN COLORADO

BOOKS BY RICHARD M. PEARL

Colorado Rocks, Minerals, Fossils
Colorado Gem Trails and Mineral Guide
America's Mountain: Pikes Peak and the Pikes Peak Region
American Gem Trails
Gems, Minerals, Crystals, and Ores: The Collectors' Encyclopedia
Successful Mineral Collecting and Prospecting
How to Know the Minerals and Rocks
1001 Questions Answered About the Mineral Kingdom
1001 Questions Answered About Earth Science
Geology: Physical and Historical
Rocks and Minerals
Wonders of Gems
Wonders of Rocks and Minerals
Nature As Sculptor: A Geologic Interpretation of Colorado Scenery
Guide to Geologic Literature
Popular Gemology
Mineral Collectors Handbook
Colorado Gem Trails
The Art of Gem Cutting (With Dr. H. C. Dake)
Seven Keys to the Rocky Mountains
Geology: An Introduction to Principles of Physical and Historical Geology
Geology Simplified
The Wonder World of Metals
Gem Identification Simplified

To
HARVEY C. MARKMAN
Curator Emeritus
Department of Geology and Paleontology
Denver Museum of Natural History

In deep respect and appreciation
for his dedication of a lifetime
to the advancement of the earth sciences
in Colorado

Preface

Rocks, minerals, and fossils are — as expressed by Ray Smith Bassler — usually thought of as the objects geologists study. Rocks furnish fundamental information about the composition and history of the earth. Minerals are the expression of these objects as chemicals and their artistic representations as crystals and gems. Fossils reveal the story of ancient life, its environment, and its development.

Geology as a natural science is, however, an even broader subject, because it records the entire history of the earth and its plant and animal inhabitants ever since the origin of our planet from the sun. Thus, geology looks to astronomy, physics, chemistry, botany, and zoology for their contribution in deciphering this complex history. Its discoveries lead directly into many other physical, biological, and social sciences.

The purpose of this book is to present the rocks, minerals, and fossils of Colorado in that order and in a form most useful to the student and teacher, collector and hobbyist, and general reader. It is designed to fill in an area still uncovered by the many other publications that have already been written about Colorado geology. These books and articles are freely mentioned as sources of the more detailed information that has made my book possible. Clear definitions and basic principles are given for the subjects treated here, and the rock, mineral, and fossil resources of Colorado are used as examples wherever possible.

Exploring Rocks, Minerals, Fossils in Colorado can serve as a textbook on geology, petrology, mineralogy, or paleontology. Therefore, only passing mention is made, where desirable, of detailed principles and of examples occurring outside Colorado. Within such a framework, I have tried to place the contents of this book in order to survey the earth-science aspects of nature in Colorado in an interesting and helpful manner.

The references given — many are annotated for your guidance — are of three main types: (1) Those that give credit or place responsibility for specific information; (2) those that refer to publications that, like this book, are summaries, surveys, or compilations of material otherwise widely scattered and not easily accessible; (3) those that indicate basic sources from which most later material was derived. Names of authors are given as they appear on the title pages of journal articles and books, and the names of publishers are those in current business use. Only American books are included.

This book is intended as a complement to my *Colorado Gem Trails and Mineral Guide*, also published by Sage Books. The chief purpose of the other book is to provide gem- and mineral-locality descriptions, complete with detailed directions, mileage logs, and sketch maps.

Table of Contents

			Page
Preface			6
Acknowledgments			13
Part I	Geology		
	Chapter 1	Introduction	17
	2	Geologic History of Colorado	19
	3	Sources of Information	24
Part II	Rocks		
	Chapter 4	Colorado Rocks and Formations	29
	5	Igneous Rocks	30
		Granite	30
		Pegmatite	31
		Basalt	34
		Rhyolite	35
		Perlite	35
		Pumice	36
	6	Sedimentary Rocks	38
		Sandstone	38
		Limestone	38
		Dolomite	40
		Gypsum	40
		Shale	41
		Oil Shale	41
		Clay	41
		Coal	43
		Petroleum	46
	Chapter 7	Metamorphic Rocks	48
		Schist and Gneiss	48
		Marble	48

			Page
	8	Colorado's Scenic Rocks	50
	9	Colorado Meteorites	53
	10	Tables for Identifying Colorado Rocks	61

Part III Minerals

Chapter	11	Colorado Minerals and Mining	64
		Chronology	66
		Literature	69
	12	Colorado Minerals of Particular Interest	72
	13	Type Minerals of Colorado	93
	14	Rare-Earth Minerals	102
	15	Radioactive Minerals	104
	16	Gem Minerals	109
	17	Native Metals	111
	18	Important Colorado Metals	113
		Gold	113
		Silver	113
		Copper	116
		Lead	116
		Zinc	117
		Molybdenum	119
	19	Colorado Geologic Museums	120
	20	Tables for Identifying Colorado Minerals	122
		Definition of Terms	123
		Tables	124

Part IV Fossils

Chapter	21	Fossils and Where They Are Found	150
	22	Fossil Plants	156
		Algal Limestone	156
		Higher Plants	157
		Amber	158
		Petrified Wood	159
		Uranium in Petrified Wood	159
		Petrified Forests	159
		Florissant Fossil Beds	160
		Creede Fossil Beds	163
	23	Invertebrate Fossils	166
		Mesozoic Invertebrates	167

			Page
		Tepee Buttes	168
		Insects	177
	24	Vertebrate Fossils	180
		Fossil Fish	180
	25	Fossil Reptiles	182
		Dinosaurs	182
		Morrison Dinosaurs	185
		Garden Park Dinosaurs	188
		Southeastern Colorado Dinosaur Footprints	189
		Middle Park Dinosaurs	190
		Dinosaur National Monument	190
		Colorado National Monument	193
		Literature on Dinosaurs	194
		Uranium in Fossil Bones	195
		Marine Reptiles	195
Chapter	26	Fossil Birds	197
	27	Fossil Mammals	199
		Carnivores	199
		Ungulates	200
		Perissodactyls	201
		Pawnee Buttes	203
		Horses	205
		Artiodactyls	207
		Mammoths and Mastodons	207
		Ground Sloths	208
		Rodents	209
		Maps and Literature on Mammals	209

Table of Illustrations

I: Maps, Tables, and Drawings

	Page
Geologic Time Scale	16
Geologic Features of Colorado	20
Geologic Map of Colorado	21
Granite Quarries in Colorado	28
Pegmatite Deposits in Central Colorado	32
Perlite Deposits in Colorado	34
Pumice Deposits in Colorado	35
Sandstone Quarries in Colorado	40
Limestone Quarries in Colorado	40
Oil-Shale Deposits in Colorado	42
Clay Deposits in Colorado	44
Coal Deposits in Colorado	44
Marble Quarries in Colorado	49
Rare-Earth Deposits in Colorado	103
Uranium Deposits in Central Colorado	106
Uranium Deposits in Western Colorado	107
Gold Districts in Colorado	114
Silver Districts in Colorado	115
Copper Districts in Colorado	117
Lead and Zinc Districts in Colorado	118
Map of Florissant Fossil Beds	161
Florissant Insect Fossils	178
Part of Dinosaur National Monument	191
Colorado National Monument	192
Ice-Age Vertebrate Fossil Localities in Colorado	200
Ice-Age Fossil Localities in Denver Area	
North of Colfax Avenue	210
South of Colfax Avenue	211

II: Photographs

Page

Pikes Peak (granite) flanked by uptilted sedimentary
rocks   81
Rhyolite, associated with garnet and other gems,
Arkansas Valley   82
Graphic granite, most characteristic aspect of
Colorado pegmatite   82
Granite, the most common igneous rock in Colorado   83
Basalt, the porous variety called scoria, Oak Creek   83
Shale, the most abundant sedimentary rock in Colorado   84
Sandstone, consisting mostly of quartz, is familiar
in Colorado   84
Upturned sedimentary rocks in the foothills
at Morrison   85
Leopard rock, a curious white sandstone   86
Oil-shale deposit at Rifle   87
Pumice, a light volcanic rock   88
Gneiss, a typical metamorphic rock in Colorado   88
Crinoid of Mississippian age   169
Baculite ammonite of Cretaceous age   169
World's largest petrified tree stump   170
Brachiopod of Devonian age   171
Tribolite of Cambrian age   171
Ammonite of Cretaceous age   172
Algal reef with layers of pyrite   172
The newest mounted dinosaur from Colorado   173
Quarry Visitor Center at Dinosaur National Monument   174
Revealing dinosaur vertebra at Dinosaur
National Monument   174
Professor Cope's *Antrodemus* dinosaur   175
Mastodon molar teeth of Pleistocene age   176

Acknowledgments

Personal thanks for helpful advice and information are due the following:

Helen A. Anderson, Saguache County Museum

George W. Bain, Amherst College Museum

Billy Boyles, Secretary-Treasurer, Longmont Museum, Inc.

Mrs. Charles L. Coleman, Jr., Secretary of the Board, Saguache County Museum

Bruce F. Curtis, Professor of Geology, University of Colorado

Mary Dawson, Assistant Curator, Section of Vertebrate Fossils, Carnegie Museum

Denver Convention and Tourist Bureau

David H. Dunkle, Associate Curator, Division of Vertebrate Paleontology, U. S. National Museum

Edwin B. Eckel, U. S. Geological Survey

William A. Fischer, Professor of Geology, Colorado College

C. Lewis Gazin, Curator, Division of Vertebrate Paleontology, U. S. National Museum

F. Geck

Don P. Greene, Curator, Adams State College Museum

Henry L. Gresham, Ward's Natural Science Establishment

Mrs. Ann Huntington, Museum Secretary, Stovall Museum, University of Oklahoma

Glenn L. Jepsen, Curator of Paleontology, Princeton Musem of Natural History

James E. Jones, Acting Superintendent, Dinosaur National Monument

L. B. Kellum, Museum of Paleontology, University of Michigan

A. S. Konselman, Project Coordinator, U. S. Bureau of Mines

William H. Kerns, Project Coordinator, Mineral Resources

Wann Langston, Jr., Texas Memorial Museum, University of Texas

John H. Lewis, Assistant Professor of Geology, Colorado College

13

Harvey C. Markman, Curator Emeritus, Department of Geology and Paleontology, Denver Museum of Natural History

Catherine McGeary, Staff Secretary, Museum of Comparative Zoology, Harvard University

Richard W. Moyle, Assistant Professor of Geology, Western State College Museum

A. Stuart Northrup, Professor of Geology, University of New Mexico

Harry W. Oborne, Consulting Geologist, Colorado Springs

John H. Ostrom, Yale Peabody Museum

Mrs. Mignon Wardell Pearl, for maps

John M. Rensberger, Museum of Paleontology, University of California

Horace G. Richards, Academy of Natural Sciences (Philadelphia)

L. S. Russell, Chief Biologist, Royal Ontario Museum

C. Bertrand Schultz, Director, University of Nebraska State Museum

H. G. Smiley, Fort Collins Museum

Mrs. Bernice Y. Smith, Museum Attendant, Museum of Earth Sciences, University of Utah

Fred L. Smith, Director of Research, Colorado School of Mines Research Foundation, Inc.

David Techter, Assistant in Fossil Vertebrates, Chicago Museum of Natural History

Dillwyn Thomas, Art Editor, *The Explorer*, Cleveland Museum of Natural History

U. S. Bureau of Mines

Robert D. Wilfley, Mining Geologist

Clifford P. Wilson, Acting Director, National Museum of Canada

Dorothy Wither, Tread of Pioneers Museum, Steamboat Springs

H. O. Wood, Curator, Canon City Museum Association

Librarians and staff at Charles Leaming Tutt Library, Colorado College; Denver Public Library; Norlin Library, University of Colorado

Mrs. Marcella Winters, Mrs. Anna Louise Lyons, Mrs. Jessica Warner, for typing

14

Part I

GEOLOGY

GEOLOGIC TIME SCALE

Era	Period	Epoch	Duration in Millions of Years	Began Millions of Years Ago
Cenozoic	Quaternary	Recent	(Late archaeologic and historic time)	
		Pleistocene	1	1
	Tertiary	Pliocene	12	13
		Miocene	12	25
		Oligocene	11	36
		Eocene	22	58
		Paleocene	5	63
Mesozoic	Cretaceous		72	135
	Jurassic		46	181
	Triassic		49	230
Paleozoic	Permian		50	280
	Carboniferous:			
	Pennsylvanian and		30	310
	Mississippian		35	345
	Devonian		60	405
	Silurian		20	425
	Ordovician		75	500
	Cambrian		100	600
Pre-Cambrian:				
Proterozoic and Archeozoic			900 (Undetermined)	1,500

Chapter 1

Introduction

Rocks, minerals, and fossils — the subjects of this book — are properly discussed together because they occur together in the earth. At one time, they were all called fossils, from the Latin word *fossilis,* which was derived from the past participle *(fossus)* of the verb *fodere,* meaning "to dig." Today, we regard only the remains or indications of ancient life, either plant or animal, as *fossils; minerals* are the inorganic chemicals of natural origin; and *rocks* — which may consist of one or many minerals, and which may contain fossils in them — are considered the structural materials of which the earth is built. Some rocks are not made of minerals; examples of this kind of rock are obsidian (which is volcanic glass) and coal (which is of plant origin, hence organic). These points are brought out further in the book.

The wide range of earth substances found in Colorado is the result of the diversity of geology for which the state is noted. Here, first of all, is a strictly political unit that takes on the physical features of each of the seven states that border it — only Missouri and Tennessee, with eight, have more neighbors. Its five physiographic provinces — only Alaska has more — are different natural units, each of which has its own characteristics of rocks and topography.

A brief outline of these five natural regions seems in order. The eastern north-south zone of Colorado belongs to the *Great Plains,* which enter from Kansas and Nebraska at a minimum altitude of 3,350 feet along the Arkansas River and reach 6,500

17

feet at the foothills of the mountains. The central zone embraces the *Southern Rocky Mountains,* with intervening valleys and open spaces known as parks; the highest part is the Sawatch Range, where Mount Elbert attains an altitude of 14,431 feet. The western zone is the *Colorado Plateau,* from 5,000 to 11,000 feet above sea level. The two smaller provinces are the south-eastern end of the *Middle Rocky Mountains* and the southern portion of the *Wyoming Basin,* a rolling prairie interrupted by streams and isolated mountains.

The kinds of rocks and minerals in the various parts of Colorado depend upon the nature of the geology — such as placer deposit, lava flow, contact metamorphism, and other types. The kinds of fossils, however, depend also upon the age of the sedimentary rocks, the only type in which they customarily occur. This is because fossils reflect the evolutionary history of life on the earth, and once an animal or plant becomes extinct or has evolved to a different species, it never returns to its previous form.

Principally, for its use in referring to the occurrence of Colorado fossils, but also for dating rock formations described in this book, the condensed, standard *geologic time scale* is printed here. You will notice that the *eras* — the major "chapters" in earth history — are divided into *periods,* and the periods are divided into *epochs.* The first two eras are together termed Precambrian time. The Mississippian and Pennsylvanian Periods are together called Carboniferous time, because that was the age of the great coal making in other parts of the world. The number of years given in the time scale has been calculated from the radioactive disintegration of uranium into lead and by other similar means. Not all the eras, periods, and epochs are known in Colorado rocks, although a substantial proportion of total geologic time is represented by the rocks of this state. The missing intervals are accounted for by times of erosion instead of deposition, or else the fossils by which dating can be accomplished are missing, or the rocks have been too strongly altered (metamorphosed) to be identified or too deeply buried to be accessible.

Chapter 2

Geologic History of Colorado

The story of Colorado geology goes back far into Precambrian time, when the part of the world that is now Colorado began to grow by accretion upon the core of the North American continent. This core is known now as the Canadian Shield, occupying more than 2 million square miles in eastern Canada, some of northeastern United States, and Greenland. The shield was enlarged by the extremely slow addition of rock from the surrounding ocean floors, which became mountain belts and were then added to the continental crust. In succession, one belt after another of younger rock became part of an expanding continent, reaching the position of Colorado about 1,250-1,450 million years ago.

Since the first appearance of the most ancient Colorado rocks that can be recognized within the present borders of the state, a succession of mountain ranges has come into being here. As each was eroded away, the sea advanced into the great interior of the continent, and then retreated when the earth rose again. The geologic history of Colorado is a record of the sediment that has been deposited in these seaways and upon the land (mostly by streams) and then turned to sedimentary rock, together with the igneous rock that has come up into them from within the crust, or has flowed out upon the surface as lava. With the changing physical structure, the climate and vegetation have also changed, and the life of land and sea has evolved correspondingly.

The oldest rocks in Colorado are metamorphic rocks, which

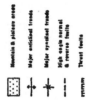

Geologic Features of Colorado. *Bruce F. Curtis, Guide to the Geology of Colorado, Geological Society of America, 1960.*

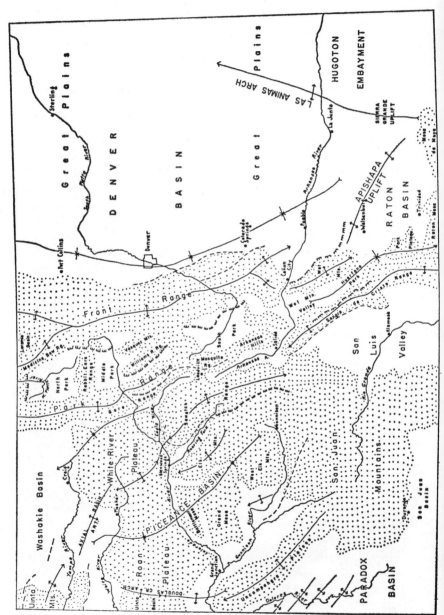

Geologic Map of Colorado. *Bruce F. Curtis, Guide to the Geology of Colorado, Geological Society of America, 1960.*

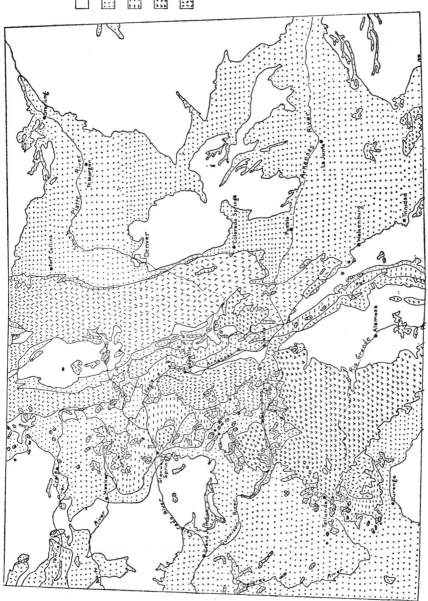

have altered from still-older sedimentary rocks and probably some lava. These were later invaded by large bodies of molten magma, which cooled to become intrusive igneous rocks (mostly granite), replacing and altering some of the metamorphic rock. These events took place during the Precambrian eras, when land life was nonexistent and marine life consisted mostly of lowly forms. Some sedimentary rock, only slightly metamorphosed, also represents this geologic age.

Extensive erosion at the close of Precambrian time was followed by the entrance of the sea, which came and went repeatedly during the Paleozoic Era. An ever-changing relation between land and water marked the long geologic history of the state at that time. While the age of invertebrates yielded to the age of fishes, and this in turn was replaced by the age of amphibians, plants of evolving types spread across the land. A mountainous terrain rose above the seas in three areas of the state, especially toward the end of the era, when the Ancestral Rockies, or Colorado Mountains — the product of the Appalachian orogeny—emerged as one of the most conspicuous features in Colorado's geologic past.

A somewhat similar story, differing in detail, is that of the Mesozoic Era, which reached a climax of marine sedimentation and then terminated in the Laramide orogeny, which created the present generation of Rocky Mountains. This was the age of reptiles and medieval plants.

The general trend of these Southern Rockies was influenced by a similar trend of the earlier Ancestral Rockies, or Colorado Mountains, lying northwestward across the state. Athwart it is a transverse zone that runs northeastward and is marked by bodies of intruded igneous rock and by almost all (except Cripple Creek) of the main mineral deposits of Colorado. This transverse zone is known as the Mineral Belt (see page 64).

The huge intrusions of igneous rock continued into the Cenozoic Era, which also saw the development in Colorado of widespread volcanic activity. With the final disappearance of the sea, the sedimentary rocks are all of continental (nonmarine) origin, deposited in basins bordering the mountain ranges, which have been uplifted and eroded in several stages. This, the age of mammals, ended with the Ice Age — but of course it has not really ended, even though man has become a geologic factor, possibly of profound significance.

A popular outline of Colorado geology is *Geology and Natural Resources of Colorado,* by Russell D. George (University of Colorado, Boulder, 1927).

An older, popular survey of Colorado geology, excellent but now rather out of date, was written by R. C. Hills for Volume II of *History of the State of Colorado,* by Frank Hall (Rocky Mountain Historical Company, Chicago, 1890).

General, technical summaries of Colorado geology are given by Bruce F. Curtis in *Guide to the Geology of Colorado* (Geological Society of America *et al,* Denver, 1960, pages 1-8) and by John W. Vanderwilt in *Mineral Resources of Colorado* (State Mineral Resources Board, Denver, 1947, pages 1-14).

Summarized information on Colorado geology will be found in *Colorado,* Guidebook 19: Excursion C-1, International Geological Congress 16th session, Washington, 1932.

The physical framework of Colorado is thoroughly treated on a regional basis in *Structural Geology of North America,* by A. J. Eardley (Harper & Row, Publishers, Inc., New York, 2nd edition, 1962), and less extensively in *The Evolution of North America,* by Philip B. King (Princeton University Press, Princeton, New Jersey, 1959).

A general survey of geology is *Geology; An Introduction to Principles of Physical and Historical Geology,* by Richard M. Pearl (Number 13 in the College Outline Series of Barnes & Noble, Inc., New York, 3rd edition, 1963). This book is keyed to 22 standard textbooks, which together furnish a great deal of information on the broad subject of the earth sciences.

The detailed "Bibliographies of North American Geology," published as Bulletins of the U. S. Geological Survey, carry the literature back to 1785.

Chapter 3

Sources of Information

Information about Colorado rocks, minerals, and fossils is available from a number of federal and state agencies and other sources. Only a selected list is presented here, dealing with general rather than specialized matters. Detailed guides to services and sources of information are listed in References 1-3. You might first look into the references mentioned, under the subjects discussed in the book, before seeking aid elsewhere.

Mineral specimens found in Colorado will be identified free to residents ($1.00 per sample for out-of-state residents) if sent to the Prospector Mineral Identification Service, P.O. Box 112, Golden 80401.

Information on Colorado mines and mining activity can be obtained from the Colorado Bureau of Mines, 215 Columbine Building, 1845 Sherman Street, Denver 80203. The Bureau's splendid museum, formerly under the auspices of the State Historical Society of Colorado, is operated in Golden by the Colorado School of Mines.

Information on locating mining claims is given by the Colorado Bureau of Mines (above) and the U. S. Bureau of Land Management, 14002 Federal Building, 20th and Stout Streets, Denver 80202.

Publications on Colorado geology, rocks, minerals, and fossils are issued by the U. S. Geological Survey. Most of these are sold by the Superintendent of Documents, Government Printing Office, Washington, D. C. 20025. They are also sold at 1012 Federal Building, Denver (above). The free publications are available from the U. S. Geological Survey, Washington 25, D. C. Maps of Colorado mineral interest are sold by the U. S. Geological Survey, Washington, D. C. 20025; 1012 Federal Building, 20th and Stout Streets, Denver 80202; and Building 41, Federal Center, Denver 80225. Catalogs of publications and maps are distributed free by the U. S. Geological Survey (above addresses).

Publications on Colorado minerals and mines are issued by the U. S. Bureau of Mines. Most of them are sold by the Superintendent of Documents, Government Printing Office, Washington 25, D. C. Monthly lists are distributed free, and indexed catalogs are sold. The Bureau's Denver office is Building 20, Federal Center, 80225.

The principal libraries containing books on these subjects are the Denver Public Library (Science and Engineering Department); U. S. Geological Survey Library, Building 25, Federal Center, Denver 25; University of Colorado (Norlin Library, Geology Department Library), Boulder; Colorado School of Mines (Arthur Lakes Library), Golden; Colorado State University, Fort Collins; Colorado College (Charles Leaming Tutt Library), Colorado Springs.

REFERENCES

Ref. 1.   Colorado School of Mines Mineral Industries Bulletin, vol. 3, no. 2, 1960; vol. 5, no. 6, 1962; vol. 6, no. 4, 1963.
Ref. 2.   Directory of Geological Material in North America, American Geological Institute, Washington, 1957.
Ref. 3.   Guide to Geologic Literature, by Richard M. Pearl, McGraw-Hill Book Company, New York, 1951.

Part II

ROCKS

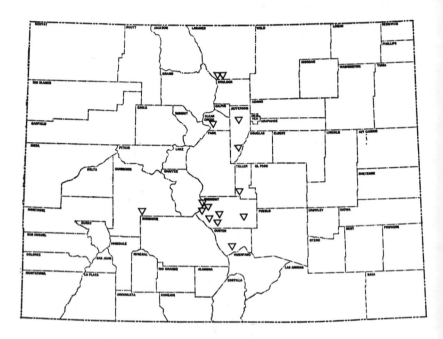

Granite Quarries in Colorado. *Colorado School of Mines.*

Chapter 4

Colorado Rocks and Formations

Throughout Colorado's long geologic history, measured in hundreds of millions of years (see Chapter 2), *igneous rocks* have solidified from molten rock both beneath and upon the surface of the earth. *Sedimentary rocks* have accumulated from broken fragments of older rock; they have also precipitated from solutions saturated with dissolved mineral matter. *Metamorphic rocks* have developed by the action of heat, pressure, and fluids on igneous and sedimentary rocks. Rocks have thus formed and been deformed; they have originated in one place and been carried elsewhere; their evolution is a fascinating and essential part of the natural history of Colorado.

These rocks are now present as solid bodies known as *formations*. Most formations are layers of sedimentary rock deposited under similar conditions during a rather restricted interval of time. Each is given a geographic name according to the type locality where it was first described. Thus, we have the Lyons sandstone, named after the town of Lyons; the Fountain formation, named after Fountain Creek; and the Carlile formation, named after Carlile Spring and Carlile Station. Nevertheless, bodies of igneous or metamorphic rock, not in layers, are also given formation names to show that they are individual units that can be described and mapped.

A total of 426 geographic formation names, published through 1955, have been proposed for rock units for which Colorado includes the type area or locality. Of these, 234 were acceptable to the U. S. Geological Survey in 1958. Both lists, together with the geologic ages, are given in U. S. Geological Survey Bulletin 1056-B (1959).

Chapter 5

Igneous Rocks

GRANITE

The principal igneous rock of the continents, granite is believed to form both by the cooling of molten rock at depth, and by the alteration of other rock in a solid or nearly solid state. Quartz and feldspar are the essential minerals; hornblende and mica (especially biotite) are common associated minerals; and numerous other minerals occur as minor, accessory constituents.

In 1950, granite was voted first choice for the state rock by the Colorado Mineral Society. Few of the quarries in Colorado that have produced granite have operated in recent years. Terrazzo and other decorative chips are the chief product of some of them. The Aberdeen quarry (Gunnison County) produced the granite for the State Capitol Building and State Museum Building. From the Cotopaxi quarry (Fremont County) came the stone for the Denver City and County Building. Salida granite (from Chaffee County) has been used for the spectacular Mormon Battalion Monument, in Salt Lake City. The United States Mint, in Denver, was built of granite from Masonville (Larimer County). The Platte Canyon red granite from near Buffalo Creek (Jefferson County) has been used for many buildings in Colorado, Wyoming, and Nebraska. The Post Office Building in Ogden, Utah, was made of granite from the Reece Gulch quarry, near Texas Creek (Fremont County).

Diorite is another intrusive igneous rock, darker in color and less silicic in chemical composition but otherwise similar to granite. Gabbro is still darker and less silicic, but higher in iron and magnesium content.

PEGMATITE

To mineral collectors, pegmatite is the ultimate source of the best specimens. This coarse- and variable-grained rock — usually a variety of granite — is the home of large and choice crystals, fine gems, and rare minerals, many of which are seldom found anywhere else. Colorado is a noted area for pegmatites. Many of the localities described in *Colorado Gem Trails and Mineral Guide,* by Richard M. Pearl (Sage Books, Denver, 2nd edition, 1964), are pegmatite dikes and similar bodies of other shapes.

In his enumeration of noteworthy mineral localities, Edwin B. Eckel (Ref. 3) has included a large proportion of pegmatites. John B. Hanley (Ref. 4) has summarized the results of extensive investigations made by Kenneth K. Landes before the Second World War, and by himself and other geologists of the U. S. Geological Survey during the war.

The greatest number of Colorado pegmatites are situated in the Front Range, in a belt extending from the Cache La Poudre River (Larimer County) to Canon City (Fremont County), a distance of 155 miles; this belt is up to 40 miles in width. Except a few in Tertiary rock, the rest are of Precambrian age and are found in igneous and metamorphic rocks. Most of the bodies are small, but a few are as much as 1 mile long and 600 feet wide, although the commercially productive ones are usually less than one-tenth that size. A study of the zoning of pegmatites will aid the collector, as it does the miner, in finding the most desirable places in which to look for specimens. These are mainly in the core and intermediate zones.

Colorado pegmatites are usually classified according to the valuable minerals they contain. Thus, there are feldspar, beryl, mica, lithium, columbium-tantalum, and rare-earth pegmatites, although a single pegmatite may have more than one kind of zone, which may be mined at different times for different products. The most numerous pegmatites in this state are those that yield potash feldspar, generally microcline or perthite, sometimes orthoclase.

31

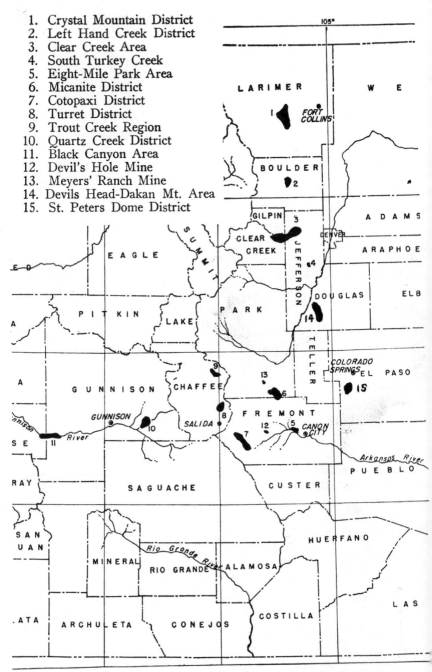

1. Crystal Mountain District
2. Left Hand Creek District
3. Clear Creek Area
4. South Turkey Creek
5. Eight-Mile Park Area
6. Micanite District
7. Cotopaxi District
8. Turret District
9. Trout Creek Region
10. Quartz Creek District
11. Black Canyon Area
12. Devil's Hole Mine
13. Meyers' Ranch Mine
14. Devils Head-Dakan Mt. Area
15. St. Peters Dome District

Pegmatite Deposits in Central Colorado. *Colorado School of Mines.*

Denman S. Galbraith (Ref. 5) has named the following 22 districts as the principal ones in Colorado:

Crystal Mountain district: Larimer County
Left-Hand Creek district: Boulder County
Jamestown district: Boulder County
Nederland district: Boulder County
Granby district: Grand County
Clear Creek district: Jefferson, Gilpin, Clear Creek Counties
South Platte district: Jefferson, Douglas Counties
Georgetown-Silver Plume district: Clear Creek County
Montezuma district: Summit, Clear Creek, Park, Grand
    Counties
Pikes Peak-Florissant district: El Paso, Teller, Douglas, Park
    Counties
Cripple Creek district: Teller, Fremont, El Paso Counties
Guffey-Micanite district: Fremont, Park Counties
Eight Mile Park district: Fremont County
Alma district: Park County
Climax district: Lake County
Cotopaxi district: Fremont County
Turret district: Chaffee County
Trout Creek district: Chaffee County
Mount Antero district: Chaffee County
Quartz Creek district: Gunnison County
Gunnison River district: Gunnison, Montrose Counties
Monarch Pass district: Fremont, Gunnison Counties

The dominant and economic minerals in these deposits are the following:

Feldspar, including microcline, orthclase, perthite, plagioclase
Quartz
Beryl
Topaz
Muscovite
Lepidolite
Columbite-tantalite
Microlite
Cerite
Euxenite

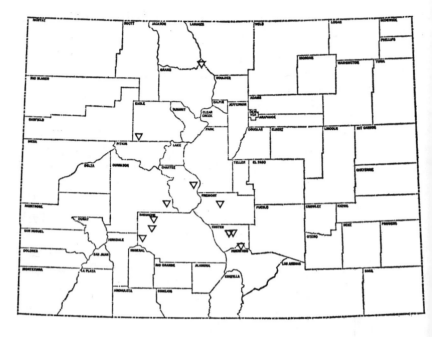

Perlite Deposits in Colorado. *Colorado School of Mines.*

Other minerals, excluding those that are present only as rare accessory minerals, include: biotite, hornblende, tourmaline, zircon, fluorite, fluorapatite, garnet, chrysoberyl, phenakite, gadolinite, bertrandite, magnetite, ilmenite, cassiterite, molybdenite, pyrite, chalcopyrite, hematite, goethite, uraninite, gummite, metatorbernite, autunite, monazite, allanite, betafite, samarskite, fergusonite, törnebohmite, bastnasite, spodumene, amblygonite, zinnwaldite, gahnite, epidote, lithiophilite-triphylite, bismuthinite, bismutite, beyerite, chlorite, triplite, scapolite, limonite, kaolinite.

BASALT

Most of Colorado's — and the world's — lava rock is basalt. Although not much used except as crushed rock for ballast, concrete aggregate, road work, and protective walls and foundations (riprap), basalt is rather widely distributed in Colorado and is

34

an important rock type in the state. The Table Mountains mineral locality at Golden (Jefferson County) is composed of basalt.

RHYOLITE

Related to basalt in being an extrusive igneous rock of volcanic (or nearly volcanic) origin is rhyolite, which is lighter in color and specific gravity (density), though usually less porous. This rock has been quarried at Castle Rock for building stone for use in Colorado Springs and elsewhere. The garnet localities at Nathrop (Chaffee County) are in rhyolite. Perlite (see below) is closely related to rhyolite and tends to occur with it. Rhyolite and similar light-colored, extrusive, igneous rocks go under the field name of felsite, because they are difficult to distinguish from one another without a microscopic study.

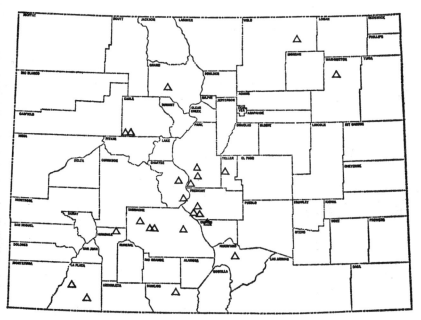

Pumice Deposits in Colorado. *Colorado School of Mines.*

PERLITE

A volcanic rock that expands into a frothy mass when heated,

35

perlite makes a lightweight aggregate serviceable in the construction industry and for other, minor purposes. Colorado is one of the leading states in its production, though far behind New Mexico. The deposits in south-central Colorado are associated with rhyolite (see page 35), to which they are related in origin. Perlite occurs in volcanic flows, dikes, and sills.

PUMICE

The explosion product of volcanoes is mostly pumice or the smaller-sized pumicite. These are porous, volcanic glass, mostly white or nearly so. The chief uses are as a lightweight aggregate and as an abrasive. Colorado has not been an important producer, but pumice occurs in a number of places in the state.

REFERENCES

Granite, Ref. 1-2; Pegmatite, Ref. 3-5; Basalt, Ref. 6-7; Rhyolite, Ref. 8-9; Perlite, Ref. 10-13; Pumice, Ref. 14-17.

Ref. 1. Colorado School of Mines Mineral Industries Bulletin, vol. 6, no. 1, 1963.

Ref. 2. Quarterly of the Colorado School of Mines, vol. 44, no. 2, 1949, p. 132-143.

Ref. 3. U. S. Geological Survey Bulletin 1114, 1961.

Ref. 4. Mineral Resources of Colorado, State Mineral Resources Board, Denver, 1947, p. 466-470.

Ref. 5. Mineral Resources of Colorado. First Sequel. State Mineral Resources Board, Denver, 1960, p. 412-420.

Ref. 6. Quarterly of the Colorado School of Mines, vol. 44, no. 2, 1949, p. 116-127.

Ref. 7. Mineral Resources of Colorado. State Mineral Resources Board, Denver, 1947, p. 252.

Ref. 8. Quarterly of the Colorado School of Mines, vol. 44, no. 2, 1949, p. 116-127.

Ref. 9. Mineral Resources of Colorado. State Mineral Resources Board, Denver, 1947, p. 252-253.

Ref. 10. Colorado Scientific Society Proceedings, vol. 15, no. 8, 1951, p. 326-331.

Ref. 11. Colorado School of Mines Mineral Industries Bulletin, vol. 4, no. 6, 1961.

Ref. 12. Quarterly of the Colorado School of Mines, vol. 44, no. 2, 1949, p. 320-338.

Ref. 13. Mineral Resources of Colorado. State Mineral Resources Board, Denver, 1947, p. 253-254.

Ref. 14. Colorado Scientific Society Proceedings, vol. 15, no. 8, 1951, p. 305-368.

Ref. 15. Quarterly of the Colorado School of Mines, vol. 44, no. 2, 1949, p. 342-347.

Ref. 16. Colorado School of Mines Mineral Industries Bulletin, vol. 3, no. 3, 1960.

Ref. 17. Mineral Resources of Colorado. State Mineral Resources Board, Denver, 1947, p. 254-255.

Chapter 6

Sedimentary Rocks

SANDSTONE

Sandstone is a sedimentary rock consisting of sand-sized grains cemented together. In commercial sandstone, used for building purposes, the grains are quartz. Conglomerate is similar but has particles of gravel size.

Of the sandstone that is quarried in Colorado, the Lyons sandstone is the most important. Best known as the building stone for the University of Colorado campus, in Boulder, it is familiar as a facing on many homes and other structures in Denver and elsewhere.

When sandstone has been metamorphosed into an extremely tough rock, it is known as quartzite. Sandstone that has merely been tightly cemented, no matter how firmly, should not be called quartzite.

LIMESTONE

The sedimentary rock consisting of the mineral calcite is called limestone. It is a product of evaporation of water containing calcium carbonate ($CaCO_3$) in solution, and much of it is of organic origin. Hot-spring and cold-spring deposits of limestone are known as travertine, which is typically porous. Cave deposits are mostly limestone.

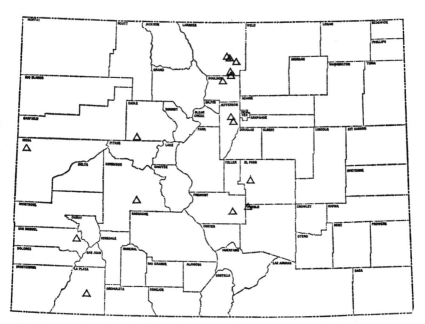

Sandstone Quarries in Colorado. *Colorado School of Mines.*

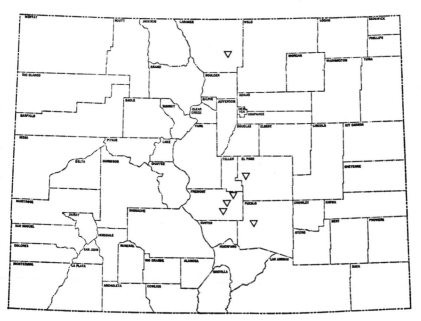

Limestone Quarries in Colorado. *Colorado School of Mines.*

Limestone has a huge commercial use. It is the basis of the cement and concrete industries and is an important raw material in metallurgy and chemical works and as a fertilizer. It is likewise a major building stone in its solid state and as terrazzo chips.

Colorado has many formations of limestone. Quarries have been opened in more than 20 counties but mainly for bulk, industrial purposes. The largest is at Monarch, in Chaffee County. Fine-quality travertine has come from Wellsville and Canon City (Fremont County); the first quarry has yielded material for the interior of the Denver City and County Building, the Department of Commerce Building, in Washington, and other large buildings throughout the country.

DOLOMITE

A carbonate rock related to limestone, dolomite is composed of the mineral dolomite, $CaMg(CO_3)_2$. It may originate as a replacement of limestone or in other ways, and it occurs in Colorado in sedimentary beds. Its commercial uses have not been applied in Colorado except as a road material near Colorado Springs.

GYPSUM

The mineral gypsum, having the chemical formula $CaSO_4 \cdot 2H_2O$, is also called gypsum when it occurs as a rock, on a large scale. It is of sedimentary origin, depositing by evaporation from enclosed basins of marine water, by concentration of subsurface water, and by replacement of limestone. Beds of shale, dolomite, and limestone are common associates of gypsum.

Fine-grained, massive bodies of gypsum are known as rock gypsum. Unconsolidated or impure material is gypsite. Compact gypsum is alabaster.

Colorado is not a noteworthy source of gypsum for the plaster and related construction industries, but deposits — some of them large — are known in at least 35 counties. The geologic age is largely Permian, Pennsylvanian, and Jurassic, when aridity prevailed in this part of the world. Three large gypsum plants are in operation at Loveland and Florence.

SHALE

This, the most abundant of all sedimentary rocks, has only incidental commercial use in the manufacture of cement and plaster. It is widespread throughout Colorado.

OIL SHALE

The largest mineral deposit in the world, excluding only the oceans, is the oil shale of northwestern Colorado and adjoining regions of Utah and Wyoming. More than 2 trillion barrels of oil are trapped in the rock of this region, nearly five times as much as the known reserves of all the rest of the globe. Given the right economic conditions, the future of shale oil is enormous.

The oil in the shale is not present as a liquid — true petroleum — but rather as an organic, solid mixture known as kerogen. This can be extracted by heating the powdered rock. The rock, furthermore, is not exactly shale but marlstone, consisting of limestone, kerogen, and other natural substances. The richest zone is the Mahogany bed of the Green River formation of Eocene age; this is in the Piceance Creek Basin, north of the Colorado River, between Glenwood Springs and Grand Junction. This sedimentary deposit was laid down in interior lakes.

A recent summary on this subject is "Oil Shale," by Charles H. Prien, in Chapter 10 of *Mineral Resources of Colorado. First Sequel* (State Mineral Resources Board, Denver, 1960). Also, "Oil Shale," by Felix C. Jaffé (Colorado School of Mines Mineral Industries Bulletin, volume 5, numbers 2-3, 1962).

CLAY

An important sedimentary rock in Colorado, clay consists of one or more of the so-called clay minerals, together with other minerals, rock fragments, and other impurities. The clay min-

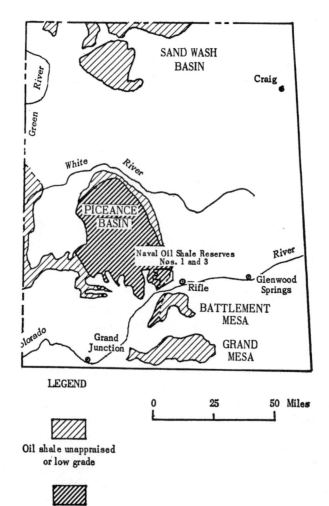

LEGEND

Oil shale unappraised
or low grade

Oil shale more than 15 feet thick,
and yielding 25 gallons of oil per
ton of shale, or more.

Oil-Shale Deposits in Colorado. *Colorado School of Mines.*

erals are complex, and their status has not been entirely worked out, but they are mostly hydrous aluminum silicates; they share in common an atomic structure that enables them to be plastic when wet and rigid when fired. Hence, clay is the basis of the ceramic industry, but many other uses are known.

COAL

The sedimentary rock called coal has had a larger part in the development of Colorado than is generally realized. Its use as fuel helped Denver to sustain itself during the early, difficult years when wood was scarce. It enabled the railroads to become successful and bind together the growing settlements of the state. It fired the mines and smelters, caused the steel mills to be built at Pueblo, generated electricity and yielded artificial gas for lighting. The decline in production since the peak was reached in 1918 seems to have been reversed, and coal mining can be depended upon to expand into the indefinite future — for it fills an industrial need nothing else can supply.

About one-third of Colorado — in 32 counties — is underlain by coal. It originates from the partial decay and burial of plants, which turn first to peat and then through the successive stages (or ranks) of lignite ("soft coal"), bituminous coal, and anthracite ("hard coal"), becoming the mineral graphite if the process should continue long enough. Colorado's coal represents a wider range of ranks than in most states — from subbituminous to anthracite — and it constitutes an extremely abundant and valuable resource. The reserves to a depth of 6,000 feet are estimated at 500 billion tons, about 40 billion tons of which can be recovered under present conditions.

The geologic age of the coal in Colorado is Cretaceous or Tertiary, having formed when the last seaway was retreating from the site of the Rocky Mountains.

Not many mineral collectors are interested in coal, because most specimens look much alike until studied carefully, when their wonderful origin and diversity becomes apparent. A particularly noteworthy variety of coal in Colorado is the coal ball found commonly in the Walsenburg field; it is a shiny, rounded

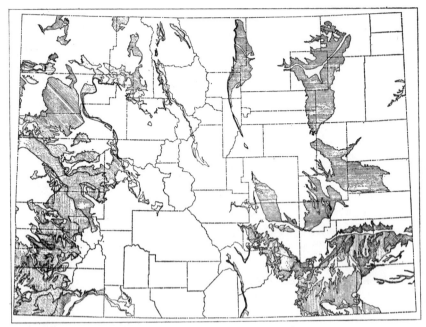

Clay Deposits in Colorado. *Colorado School of Mines.*

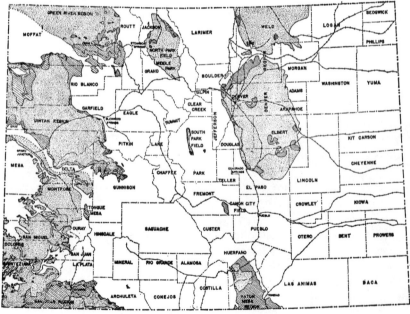

Coal Deposits in Colorado. *Parke O. Yingst and Colorado School of Mines.*

lump, a natural briquette, which owes its shape to the heating and drying effect of molten rock on a coal seam.

The coal fields of Colorado are identified by Thomas Allen as follows:

Denver region (Northern coal field)
Raton Mesa region (Southern coal field)
    Walsenburg field
    Trinidad field
Canon City field
South Park field
North Park-Middle Park field
Green River region (Yampa field)
Uinta region
    Danforth Hills field
    Lower White River field
    Grand Hogback field
    Carbondale field
    Crested Butte field
    Somerset field
    Grand Mesa field
    Book Cliff field
San Juan River region
    Durange field
    Pagosa Junction field
    Tongue Mesa field

A good deal has been written about Colorado coal. The chief general references of recent date are as follows:

"Coal Resources of Colorado," by E. R. Landis, U. S. Geological Survey Bulletin 1072-C, 1959.

"Coal," by Parke O. Yingst, in "Mineral Resources of Colorado. First Sequel," State Mineral Resources Board, Denver, 1960, p. 465-486.

"Coal Resources of Colorado," by Parke O. Yingst, "Colorado School of Mines Mineral Industries Bulletin," vol. 3, no. 5, 1960.

Petroleum, though a liquid, is a mineral resource of the utmost importance. So is the natural gas that occurs with it and is dissolved in it. These two substances, presumably of organic origin, are found typically in sedimentary rocks of all geologic ages. Hence, it seems proper to mention them briefly in this chapter.

Colorado became the second state to produce oil from wells drilled for the purpose, beginning in 1862 near Canon City. The Florence field became the first in Colorado in 1876. Numerous fields have been opened up since then, the most prolific being Rangely, which was worked as early as 1902 but did not become a major field until deep wells were drilled in 1933; currently, it ranks 11th in the entire nation and is 17th in total yield since the industry began in 1859. The reserves in Colorado, by present methods of recovery, are estimated (as of 1963) to be 388 million barrels of oil and 2,216,818 million cubic feet of gas.

The principal oil and gas fields of Colorado are included in the following regions:

> Denver Basin
> Canon City Embayment
> Las Animas Arch
> Trinidad, or Raton, Basin
> North and Middle Park
> Green River Basin
> Uinta Basin
> Paradox Basin
> San Juan Basin

The bibliography on Colorado petroleum is very extensive. A good, recent summary by Francis M. Van Tuyl and others occupies Part 4 of *Mineral Resources of Colorado. First Sequel* (State Mineral Resources Board, Denver, 1960).

REFERENCES

Sandstone, Ref. 1-3; Limestone, Ref. 4-7; Dolomite, Ref. 8-9; Gypsum, Ref. 10-13; Clay, Ref. 14-16.

Ref. 1. Colorado School of Mines Mineral Industries Bulletin, vol. 6, no. 1, 1963.

Ref. 2. Quarterly of the Colorado School of Mines, vol. 44, no. 2, 1949, p. 132-143.

Ref. 3. Mineral Resources of Colorado, State Mineral Resources Board, Denver, 1947, p. 249-250.

Ref. 4. Colorado School of Mines Mineral Industries Bulletin, vol. 6, no. 1, 1963.

Ref. 5. Colorado School of Mines Mineral Industries Bulletin, vol. 3, no. 1, 1960.

Ref. 6. Quarterly of the Colorado School of Mines, vol. 44, no. 2, 1949, p. 254-274, 459-464.

Ref. 7. Mineral Resources of Colorado, State Mineral Resources Board, Denver, 1947, p. 244-247.

Ref. 8. Quarterly of the Colorado School of Mines, vol. 44, no. 2, 1949, p. 144-149.

Ref. 9 Mineral Resources of Colorado, State Mineral Resources Board, Denver, 1947, p. 240.

Ref. 10. Colorado School of Mines Mineral Industries Bulletin, vol. 6, no. 2, 1963.

Ref. 11. Quarterly of the Colorado School of Mines, vol. 44, no. 2, 1949, p. 226-238.

Ref. 12. Mineral Resources of Colorado, State Mineral Resources Board, Denver, 1947, p. 242-244.

Ref. 13. U. S. Geological Survey Bulletin 697, 1920.

Ref. 14. Colorado School of Mines. Mineral Industries Bulletin, vol. 4, no. 4, 1961.

Ref. 15. Quarterly of the Colorado School of Mines, vol. 44, no. 2, 1949, p. 89-109.

Ref. 16. Mineral Resources of Colorado. State Mineral Resources Board, Denver, 1947, p. 232-240.

Chapter 7

Metamorphic Rocks

SCHIST AND GNEISS

These are metamorphic rocks, rather widespread among the older (Precambrian) rocks in Colorado. They differ chiefly in the coarseness of the banded texture, gneiss consisting of wider bands than schist. Gneiss is also characterized by alternate bands of different minerals, whereas schist may be composed of one mineral in repeated sheets or layers. The presence of feldspar is also typical of gneiss. These rocks are exposed in deep canyons and in the cores of the mountain ranges.

MARBLE

When limestone and dolomite are changed into metamorphic rock by heat, pressure, and fluids, they become marble. Chemical impurities cause a variation in color and pattern, which may become extremely beautiful, although the superb Yule marble of Colorado is snow white. Terrazzo and other chips are now a common product, and other marble is quarried for crushing.

From the Yule Colorado Marble Company deposit, near Marble, in Gunnison County, came the stone for the Lincoln Memorial, the first Tomb of the Unknown Soldier (a 56-ton block), the municipal buildings in New York and San Francisco, the Field Building in Chicago, the Denver Post Office Building, Custom House Building, and Federal Reserve Bank, and other outstanding buildings in Denver and elsewhere. The Mormon Auditorium at Independence, Missouri, was constructed of white marble from a different quarry not far away. The Yule marble has been described in detail by John W. Vanderwilt (Ref. 1) and B. B. Bartholomew (Ref. 2).

REFERENCES

Ref. 1. U. S. Geological Survey Bulletin 884, 1937, p| 163-173.
Ref. 2. 1940 Mining Year Book, Colorado Mining Association, 1940, p. 42-43.
Ref. 3. Colorado School of Mines Mineral Industries Bulletin, vol. 6, no. 1, 1963.
Ref. 4. Quarterly of the Colorado School of Mines, vol. 44, no. 2, 1949, p. 280-297.

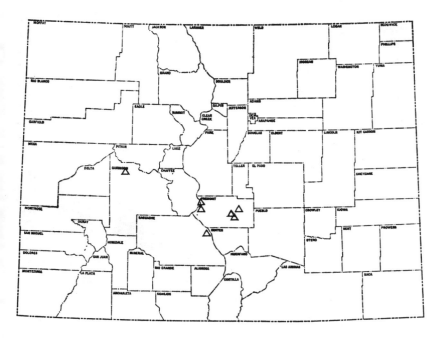

Marble Quarries in Colorado. *Colorado School of Mines.*

49

Chapter 8

Colorado's Scenic Rocks

This title, taken from one used by the Virginia Division of Mineral Resources, brings vividly to mind the role played by rock in the development of the scenery for which Colorado is so famous. For scenery is primarily rock. The details of the scenery depend on the kinds of rock of which the landscape is composed, the manner in which these rocks have been assembled in making the architectural features we see, and the geologic processes that have been at work within and upon them. Vegetation, of course, adds considerably to the aesthetic value of scenery, but it grows only from the soil that develops as the rock breaks down and decays under the influence of the atmosphere. For billions of years, it should be remembered, there was no vegetation on the land.

The nature of the Colorado scenery varies according to the physiographic divisions of the state, as discussed in Chapter 1. Each section has its own kind of scenery. The rocks and geologic agents have produced vast plains on one side of Colorado, lofty plateaus on another side, and rugged mountains in between. The two national parks and five national monuments are either principally geologic in significance or, as is true of the southwestern cliff dwellings, their geology has been the fundamental factor in bring about their utilization by man.

Three factors combine in countless ways to yield the diversified scenery of today: *structure* (the kinds of rock and their relative positions), *process* (the geologic agents at work), and *stage* (the point reached in the cycle through which these agents pro-

50

gress). The same agents may form quite unlike scenery in two different kinds of rock. Thus, the soft shale of the mountains is readily worn into valleys by streams, while the hard quartzite resists mightily and remains as rugged spires even where glaciers have been most active.

Volcanic rocks are responsible for much of Colorado's splendid scenery. Although no live volcano exists in Colorado today, the evidence of past activity is clear enough. Extinct volcanoes, such as Specimen Mountain — on the Continental Divide, at an altitude of 12,482 feet, in Rocky Mountain Park — stand as proof. Cripple Creek occupies the throat of an old volcano that burst through the Pikes Peak granite during the Miocene Epoch. Some of the Colorado hot springs suggest former volcanism.

The most widespread volcanic material in Colorado is the vast amount of basalt and other lava rock that occupies much of the state. The towering summits of the San Juan Mountains — the Switzerland of America — are carved from successive outpourings of such rock, which welled out of the earth and piled up to heights of thousands of feet. Other lava flows spread eastward from the Rocky Mountains almost to the Kansas border during the Ice Age.

Similar, heat-generated processes, but confined within the crust, give rise to the *intrusive* igneous rocks, of which granite is the most familiar. Granite is exposed in the cores of Colorado's mountains. Other bodies of igneous rock (called *stocks)* make up the Spanish Peaks, west of Walsenburg and Trinidad, described by Spanish conquistadores as "the great isolated double mountain situated at the northernmost limits of the Empire." Over 500 enormous walls of rock (*dikes)* radiate as much as 15 miles in every direction from both mountains.

Within the earth, but near the surface, are the scenic caverns of Colorado. Practically all of them are in limestone, a few in marble. Some, such as the Cave of the Winds, at Manitou Springs, are well advertised, while others are little known, and many doubtless remain to be explored.

Uplifted sedimentary rock is responsible for much fine scenery in Colorado. The Garden of the Gods, at Colorado Springs, the Park of the Red Rocks, near Denver, and the Grand Hogback, which runs spectacularly across the western part of the state, are but a few examples of what can be done by determination and patience. Nature has plenty of both.

Metamorphic rock — the third major type of rock — is present in numerous Colorado canyons and gorges in the heart of the mountains. The Needle Mountains, in the San Juans, are built of the hardest and most enduring kind of metamorphic rock: quartzite.

A brief elementary survey of the geology of Colorado, with emphasis on the topography and the causes of the more conspicuous aspects of the scenery, is *Nature As Sculptor; A Geologic Interpretation of Colorado Scenery,* by Richard M. Pearl (Denver Museum of Natural History, 3rd edition, 1956).

Broader coverage of the same subjects, in which Colorado is well represented, is given in the following regional geomorphologies of the United States:

This Sculptured Earth: The Landscape of America, by John A. Shimer (Columbia University Press, New York, 1959).

Physiography of the United States, by Frederic B. Loomis (Doubleday and Company, Inc., New York, 1937).

The Physiographic Provinces of North America, by Wallace W. Atwood (Ginn and Company, Boston, 1940).

Physiography of the United States, by Charles B. Hunt (W. H. Freeman and Company, San Francisco, 1967).

Physiography of Western United States, by Nevin M. Fenneman (McGraw-Hill Book Company, New York, 1931).

Regional Geomorphology of the United States, by William D. Thornbury (John Wiley and Sons, Inc., New York, 1965).

The various aspects of the mountains of Colorado are covered in *Guide to the Colorado Mountains,* by Robert M. Ormes (Sage Books, Denver, 4th edition, 1963).

Chapter 9

Colorado Meteorites

Colorado ranks next only to Texas and Kansas in the number of meteorites that have been found. It is one of the leading areas of its size in the world. These rocks — the only things that man has yet been able to see and touch and analyze that have their origin outside the earth — are among the most interesting of all natural objects. In recent years, they have taken on a new significance as samples of outer space — probably bits of another planet — and they are being studied intensively. Any newly found specimen has both a scientific and a commercial value and should be reported promptly.

The large number of known Colorado meteorites (as of adjacent states, too, for that matter) is most directly the result of many years of work in this region by Harvey H. Nininger (now of Sedona, Arizona), who lectured extensively on the subject and traveled widely into the farm and ranch country, where most of the specimens have been found. He is also largely responsible for the fine collection of meteorites in the Denver Museum of Natural History. The American Meteorite Labratory, P. O. Box 2098, Denver 1, was established by him and is a convenient place to send a small piece of any supposed meteorite for free, expert identification.

To enable you to recognize meteorites, the following information is given from literature issued by the American Meteorite Labratory and the Geological Survey of Canada.

Meteorites are not light, porous rocks. They are usually much heavier than ordinary rocks. They are often marked with shallow pits ("thumb prints") but are not hollow or porous inside. They are irregular in shape, not round, although a few are conical. The edges and corners are rounded or dulled. Meteorites are covered with a thin fusion-crust, which is due to the effect of friction in the atmosphere. This crust is soft and is generally dull black, rusting to brown upon exposure on the ground. Magnetism is an important feature of meteorites. It is due to nickel-iron, which is nearly always present and may constitute nearly the entire specimen. The metal shows bright and white when a piece is ground on an emery wheel.

To date, about 55 meteorite "finds" or "falls" are known in Colorado. The names given them indicate the place where found, usually a nearby town. A brief description of each of these meteorites, condensed from the British Museum catalog by Max H. Hey (Ref. 1), is given below. The Johnstown meteorite, which interrupted a funeral service on July 6, 1924, is the only one that was seen to fall — testifying further to Dr. Nininger's good missionary work in encouraging exploration in the field. Others of the Colorado meteorites have interesting stories, which I am assembling for the detailed technical report that I am writing about them. The historical accounts of their finding, as given here, have been taken from the Nininger catalog (Ref. 2). The coordinates given after the name are north latitude and west longitude. Records of these locations and the reported weights vary in the literature used, which includes the references mentioned and those by Brian Mason (Ref. 3) and Oliver Cummings Farrington (Ref. 4). To convert kilograms to pounds, multiply by 2.2.

Adams County 39° 58'; 103° 46'

Found in 1928 as a badly oxidized mass weighing 5.7 kilograms. Class: brecciated veined olivine-bronzite chondrite.

Akron **Washington County** 40° 09'; 103° 10'

Found in 1940 as a mass weighing 4-7 kilograms, but only 408 grams were preserved. Class: Olivine-bronzite chondrite.

Akron No. 2 **Washington County** 40° 09'; 103° 10'

Found in 1954 as a piece weighing 640 grams. This has also been called Akron No. 1, but is a much older fall. Class: Olivine-hypersthene chondrite.

Akron No. 3 Washington County 40° 09'; 103° 10'
Found in 1961 as a mass weighing 4 kilograms. Class: Olivine-hypersthene chondrite.

Alamosa Alamosa County 37° 28'; 105° 52'
Found in 1937 as a mass weighing 2.2 kilograms. After attending a lecture on meteorites by Dr. Nininger, Mrs. Wallrich, a collector of Indian relics, invited him to examine a rock pile of hundreds of specimens she had gathered. No meteorites were present; but after being told what they looked like, she found this meteorite on her next collecting trip. Class: olivine-hypersthene chondrite.

Arapahoe Cheyenne County 38° 51'; 102° 10'
Found in 1940 as a mass weighing 19.083 kilograms. Class: Black olivine-hypersthene chondrite.

Arriba Lincoln County 39° 18'; 103° 15'
Found in 1936 as three pieces, a fourth being added later, the total weighing 33.6 kilograms. Class: Polymict brecciated intermediate olivine-hypersthene chondrite.

Atwood Logan County 40° 32'; 103° 17'
Recognized in 1963 as a piece weighing 36.8 grams. Class: Olivine-hypersthene chondrite.

Bear Creek Jefferson County 39° 48'; 105° 05'
Found in 1866 as a mass weighing 227 kilograms. This has also been called Aeriotopos, Bear River, Colorado, Denver, Denver County, Jefferson, Jefferson County. Class: Fine octahedrite.

Bethune Kit Carson County 39° 18'; 102° 25'
Found in 1941. Fragments totaling 67 grams have been preserved. Three specimens were mounted on a cement wall in a rock garden.

Bishop Canyon San Miguel County 38° 05'; 109° 00'
Found in 1912 as a mass weighing 8.6 kilograms. Class: Fine octahedrite.

Briggsdale Weld County 40° 40'; 104° 19'
Recognized in 1949 as a mass weighing 2.23 kilograms. Class: Medium octahedrite.

Cope   Washington County   39° 40'; 102° 50'
Found in 1934 and recognized in 1937 as a group of 8-11 pieces totaling 12 kilograms. Class: olivine-bronzite chondrite.

Cortez   Montezuma County   37° 21'; 108° 41'
Found in 1940 as a piece weighing 0.7156 kilograms. Class: olivine-bronzite chondrite with metallic veins.

De Nova   Washington County   39° 51'; 102° 57'
Found in 1940 as a mass weighing 12.7 kilograms. Class: white veined crystallized spherical olivine-hypersthene chondrite.

Denver   Denver County   39° 47'; 104° 56'
Fell between July 11 and July 15, 1967, being recovered from the punctured roof of a warehouse as a piece weighing 230 grams. Class: Olivine-hypersthene chondrite.

Doyleville   Gunnison County   38° 25'; 106° 35'
Found in 1887 as a piece weighing 112 grams. Class: olivine-bronzite chondrite.

Eaton   Weld County   40° 31'; 104° 41'
Found May 10, 1931, after having been seen to fall between 5:00 and 5:30 P.M., but the occurrence has not been entirely accepted as authentic. The specimen, weighing 0.29547 kilograms, is a copper-alloy nugget, unique among meteorites.

Erie   Weld County   40° 02'; 105° 03'
Found just inside Boulder County in 1965 as a mass weighing 3.3 kilograms. Class: Olivine-hypersthene chondrite.

Fleming   Logan County   40° 41'; 102° 50'
Found in 1940 as a mass weighing 1.75 kilograms. A second stone was found later. Class: Polymict brecciated black olivine-bronzite chondrite.

Franceville   El Paso County   38° 49'; 104° 37'
Found in 1890 in a mass weighing 18.8 kilograms. Class: Medium octahedrite.

Fremont Butte   Washington County   40° 15'; 103° 16'
Recognized in 1963 as a mass weighing 6.647 kilograms, it was identified as a result of a joke played on a housewife in whose rock garden it was put. Class: Olivine-hypersthene chondrite.

Guffey Park County 38° 46'; 105° 31'

Found in 1907 as a pear-shaped mass weighing 310 kilo-
grams (682 pounds), the largest Colorado meteorite.
This has also been called Currant Creek and Park Creek.
Class: Nickel-rich ataxite.

Holly Prowers County 38° 04'; 102° 07'

Found in 1937 as two pieces weighing 0.2992 kilograms.
Class: Olivine-bronzite chondrite.

Holyoke Phillips County 40° 34'; 102° 18'

Found about 1933 and recognized in 1935 as a mass
weighing 5.4 kilograms. Class: Olivine-bronzite chondrite.

Horse Creek Baca County 37° 35'; 102° 46'

Found in 1937 on an Indian campsite as a piece weighing 570 grams. Class: Hexahedrite (?), pseudo-octa-
hedrite, "a unique combination of kamacite in tetrahe-
dral blocks separated by thin schreibersite lamallae"
(Nininger).

Hugo Lincoln County 39° 08'; 103° 19'

Found in 1936 as a piece weighing 80 grams. Class:
Olivine-bronzite chondrite.

Johnstown Weld County 40° 21'; 104° 54'

Fell July 6, 1924, at 4:20 P.M., after four explosions. A
total of 27 pieces were recovered, weighing 40.75 kilograms. This also has been called Elwell and Weld
County. Class: Monomict brecciated diognite (Hey),
hypersthene achondrite (Mason).

Karval Lincoln County 38° 43'; 103° 31'

Found in 1936 as an oriented, badly weather-cracked
piece weighing 1.104 kilograms. Class: Olivine-bronzite
chondrite.

Kelly Logan County 40° 28'; 103° 02'

Found in 1937 as a mass weighing 44.3 kilograms.
Class: White to gray brecciated olivine-hypersthene
chondrite.

Lafayette Boulder County 39° 59'; 105° 05'

Found before 1908 as a piece weighing 11 grams. Now
lost, perhaps used up in analysis. Class: Nickel-rich
ataxite (?).

Lincoln County   39° 22'; 103° 10'

Found in 1937 as 12 stones weighing 4.6 kilograms. Dr. Nininger believes that five different falls are represented. It has also been called Township No. 7 and Township No. 8. Class: Olivine-hypersthene chondrite.

Lost Lake   **Alamosa County**   37° 39'; 105° 44'

Found about 1931 and recognized in 1934 as a piece weighing 11 grams. Class: Chondrite.

Mesa Verde Park   **Montezuma County**   37° 10'; 108° 30'

Found in 1922 as an oxidized mass weighing 3.5 kilograms. It was in the Sun Shrine at the north end of Pipe Shrine House. Class: Medium octahedrite.

Mosca   **Alamosa County**   37° 38'; 105° 50'

Found in 1942 as a mass weighing 6.123 kilograms. Class: Olivine-hypersthene chondrite.

Mount Ouray   **Chaffee County**   38° 25'; 106° 13'

Found in 1898 as a piece weighing 910 grams at an altitude of 10,000 feet. Class: Medium octahedrite.

Newsom   **Alamosa County**   37° 36'; 105° 50'

Found in 1939 as a piece weighing 892 grams. Class: Olivine-hypersthene chondrite.

Ovid   **Sedgwick County**   40° 58'; 102° 24'

Found in 1939 as a mass of 6.169 kilograms. Class: Olivine-bronzite chondrite.

Peetz   **Logan County**   40° 58'; 103° 05'

Found in 1937 as a mass of 11.6 kilograms. Class: Black olivine-hypersthene chondrite.

Phillips County   40° 27'; 102° 23'

Found in 1935 as a mass of 1.4 kilograms. Class: Pallasite.

Rifle   **Garfield County**   39° 31'; 107° 50'

Found in 1948 as a mass of 102.7 kilograms. Class: Coarse octahedrite.

Rush Creek   **Kiowa County**   38° 37'; 102° 43'

Found in 1938 as three pieces totaling 9.3 kilograms. Class: Polymict brecciated olivine-hypersthene chondrite.

Russell Gulch **Gilpin County** 39° 45′; 105° 40′

Found in 1863, the first Colorado meteorite, as a mass of 13.2 kilograms. This has also been called Colorado and Gilpin County. Class: Fine octahedrite.

Seibert **Kit Carson County** 39° 18′; 102° 50′

Found in 1941 as a mass of 3.5 kilograms. Class: Olivine-bronzite chondrite.

Shaw **Lincoln County** 39° 32′; 103° 20′

Found in 1937 as two masses weighing 3.7 kilograms. Class: Amphoterite (Nininger), olivine-hypersthene chondrite (Mason).

Springfield **Baca County** 37° 25′; 102° 31′

Found in 1937 as two pieces weighing 3.2 kilograms. Class: Crystalline spherical olivine-hypersthene chondrite.

Sterling **Logan County** 40° 36′; 103° 11′

Found about 1900 as a piece weighing 679.5 grams. Class: Pallasite.

Stonington **Baca County** 37° 18′; 102° 12′

Found some years before 1942 as a weathered mass weighing 2.3 kilograms. Class: Veined crystalline olivine-bronzite chondrite.

Thatcher **Las Animas County** 37° 33′; 104° 09′

Found before 1963 as a piece weighing 2 grams. Class: Stone.

Tobe **Las Animas County** 37° 12′; 103° 35′

Found before 1963 as a mass weighing 5.386 kilograms. Class: Olivine-bronzite chondrite.

Ute Pass **Summit County** 39° 48′; 106° 10′

Found in 1894 as a piece weighing 0.120 kilograms. Class: Coarsest octahedrite.

Washington County 39° 42′; 103° 10′

Found in 1927 as a very fresh, disc-shaped mass weighing 5.75 kilograms, buried 12 inches deep in a wheat field. It perhaps fell in 1916. This has also been called Arickarie and Aricarie. Class: Nickel-rich ataxite.

Weldona   ꞏMorgan County   40° 21'; 103° 57'
Found in 1934 as a mass weighing 27.7 kilograms.
Class: Crystalline spherical olivine-bronzite chondrite.

Wiley   Prowers County   38° 09'; 102° 40'
Found in 1938 as a mass weighing 3.5 kilograms. Class:
Nickel-rich ataxite (Perry), microscopic octahedrite
(Nininger).

Wray   Yuma County   40° 04'; 102° 11'
Found in 1936 and recognized in 1938 as a piece weighing 0.2817 kilograms. Class: Olivine-bronzite chondrite.

REFERENCES

Ref. 1.   Catalogue of Meteorites . . . , by Max H. Hey, 3d edition, British Museum, London, 1966.

Ref. 2.   The Nininger Collection of Meteorites: A Catalog and a History, by H. H. Nininger and Addie D. Nininger, American Meteorite Museum, 1950.

Ref. 3.   Meteorites, by Brian Mason, John Wiley & Sons, Inc., New York, 1962.

Ref. 4.   Catalogue of the Meteorites of North America to January 1, 1909, by Oliver Cummings Farrington, Memoirs of the National Academy of Sciences, vol. 13, 1915.

Chapter 10

Tables for Identifying Rocks

Because of their tremendous diversity, and because many of them are varying mixtures of minerals, rocks cannot be identified as definitely as minerals can. The following table will help you recognize the most important Colorado rocks, all of which are described individually in Chapters 5-7. Keys to many rocks, together with drawings that show the parts of the identification to look for, will be found in *How to Know the Minerals and Rocks*, by Richard M. Pearl (McGraw-Hill Book Company; New American Library of World Literature, Inc).

Can be scratched by a knife blade:

Group 1. Dissolves in acid by fizzing:

Slowly   Dolomite

Rapidly   Limestone (may contain fossils)
                Marble (no fossils)

Group 2. Does not dissolve in acid by fizzing:
          Gritty feel on teeth   Shale (earthy odor when moist)

          Smooth feel on teeth   Clay (earthy odor when moist)

                                        Gypsum (not earthy odor when moist)

                                        Coal (burns)

Can scratch glass:

Group 3. Parallel arrangement of individual minerals (not as beds of layered rock):

Narrow bands, often of same mineral   Schist

Coarse bands of different minerals   Gneiss

Group 4. Not parallel arrangement of individual minerals (but may occur as beds of layered rock):

Light-colored or red rock	Granite
Medium-gray rock	Diorite
Dark-green to black rock	Gabbro
Accumulation of sand grains	Sandstone (breaks around grains)
Accumulation of gravel	Conglomerate
Fine-grained, tough	Rhyolite, Felsite (light colored)
	Basalt (dark colored)

Part III

MINERALS

Chapter 11

Colorado Minerals and Mining

Colorado is richly endowed with minerals. The 8 billion dollars' worth of new, primary wealth added to the world by the mines and wells of this state interests the economist, the industrialist, and the historian. But the hundreds of kinds of individual minerals that have come from this area are of even greater interest to the geologist and mineral hobbyist. Edwin B. Eckel (Ref. 1) has enumerated 445 Colorado mineral species, more than have been reported from any other state except California, although New Mexico has about the same number. Many more minerals already known elsewhere must still remain undiscovered, and doubtless many others await first identification in Colorado. Indeed, 42 minerals had, up to 1957, made their first appearance in this state, becoming "type specimens."

A *mineral* is defined as a natural, inorganic chemical element or compound. There are perhaps 1,600 to 2,000 mineral species but many thousands of varieties based on color, origin, or other differences, and many thousands more of duplicate names, some of which are almost impossible to fit into a satisfactory classification. *Mineral products,* or *mineral resources,* are another matter, for many of them are not simple minerals. Thus, coal, petroleum, and building stones are familiar and important mineral resources, yet these substances are usually not regarded technically as minerals. Most *rocks* (see Part II) are aggregates of two or more minerals.

The Colorado Mineral Belt corresponds to the transverse zone that cuts across the general trend of the Southern Rocky Moun-

64

tains (see page 22). Although often considered to belong only to the Front Range, where it may be called the Front Range Mineral Belt, it actually extends 250 miles from Boulder County to the western San Juan Mountains. It is characterized by a large variety of metals: gold and silver; lead, zinc and copper; molybdenum and other, less-familiar ones.

The story of Colorado minerals goes back to the distant geologic past, discussed in Part I. The history of mineral finding by man begins with the American Indian, who used earth products in religious ceremonials, for personal adornment, as weapons and tools and for decorating these objects. As shown by the discoveries of Folsom and Yuma sites, Colorado was early inhabited by the aborigine. The later history of the Indian shows his familiarity with this state, as evidenced by the spectacular cliff dwellings of the southwest, occupied by the ancestors of the Pueblo Indians of today. During the white man's time, however, Colorado was relatively deserted by the Indian, the sparse population being represented by only a few roaming Ute, Cheyenne, Arapaho, Kiowa, Sioux, Pawnee, Comanche, Shoshone, Navaho, and Jicarilla Apache. The use of gems and some other minerals by the Indian is discussed in *American Gem Trails,* by Richard M. Pearl (McGraw-Hill Book Company, New York, 1964), but Colorado receives little specific attention apart from its turquoise, which is covered more thoroughly in *Colorado Gem Trails and Mineral Guide* (Sage Books, Denver, 2d edition, 1964).

Some of the first white explorers, trappers, and guides found gold; the reputed discovery of gold near present Lake City by one of Frémont's party in 1848 seems likely enough. During the rush to California between 1849 and 1852, gold was discovered on Cherry Creek, Ralston Creek, and the South Platte River by the Russell brothers and those who accompanied them on their way west. Gold in paying amounts was first found in 1858 by the returning Russell party in the Denver area and along the front of the mountains, beginning the misnamed "Pike's Peak or Bust" gold rush. This was intensified by the discovery of lode gold — native metal in veins "in place" — on May 7, 1859, by John H. Gregory on North Clear Creek, near Black Hawk. Here began a vital mining industry, the most significant influence in the development of Colorado as a State of the Union.

The detailed history of Colorado mining, including a chronological record, is *Mining in Colorado; A History of Discovery, Development, and Production,* by Charles W. Henderson (U.S. Geological Survey Professional Paper 38, 1926). Of the many histories of Colorado, the one best regarded for its treatment of mining is *Colorado; the Centennial State* by Percy Stanley Fritz (Prentice-Hall, New York, 1941). A good summary appears in Chapters 7-8 of *The Bonanza West; The Story of the Western Mining Rushes,* by William S. Greever (University of Oklahoma Press, Norman, 1963). The story of the ghost mining towns of Colorado has never been treated better than in Muriel Sibell Wolle's now-classic *Stampede to Timberline: The Ghost Towns and Mining Camps of Colorado* (Sage Books, Denver, 1949).

CHRONOLOGY

The following chronology of Colorado mining history, modified from Henderson, covers the more significant events.

1807   James Purcell reported to Lieutenant Pike at Santa Fe that he had, in 1803, refused to lead Spaniards to an occurrence of gold on the South Platte. This report was ignored.

1848   Report of a placer-gold discovery near Lake City by a member of the Frémont party.

1849, 1850, or 1852   Placer gold found near Denver and taken away in goose quills.

1858   Important discoveries of placer gold in the Denver area, especially by the Russell and Lawrence parties.

1859   Discovery of lode gold; extension of the Pikes Peak gold rush; discoveries of gold in central Colorado.

1860   Climax and enlargement of the Pikes Peak gold rush; rush to California Gulch placer deposits at Leadville; Clark, Gruber and Company bank and mint established.

1861   Territory of Colorado organized.

1862   First oil well drilled near Canon City, the only one outside Pennsylvania.

1864   First profitable silver mine discovered, near Georgetown.

1868　Boston and Colorado Smelting Company smelter operated at Black Hawk; first use of power tools in American mining, at Silver Plume.

1870　Major railroads, Denver Pacific and Kansas Pacific, opened to Denver.

1872　Federal mining law established.

1874　San Juan mining rush begun; Colorado School of Mines opened.

1876　State of Colorado admitted; first oil field opened near Florence.

1877　Rush to mining silver-rich lead carbonate at Leadville.

1879　Discovery of silver at Aspen.

1891　Rush to Creede.

1892　Rush to Cripple Creek.

1893　Demonetization of silver, a sharp blow to silver camps.

1895　Invention of Wilfley table for ore recovery at Kokomo.

1896　Camp Bird mine relocated near Ouray by Thomas F. Walsh.

1897　Uranium minerals discovered in western Colorado.

1899　Tungsten discovered in Boulder County.

1905　Zinc mining begun in the Gilman district.

1918　Molybdenum mining begun at Climax.

1933　Price of gold raised from $20.67 to $35.00 per ounce, renewing placer mining; Colorado's largest oil field opened at depth, at Rangely.

1950　Uranium boom began, chiefly in western Colorado.

Edwin P. Eckel (Ref. 2) has prepared a summary of the valuable minerals in Colorado, emphasizing individual species rather than the metal content. About 200 of the 445 Colorado minerals are economically valuable. Of the 3 billion dollars produced in the first century of metal mining, molybdenite led the rest with 542 million dollars. Arranged by mineral, the list follows:

$542 million	Molybdenite
410	Galena
400	Native gold
295	Sphalerite

250	Calaverite
150	Krennerite
145	Cerargyrite and other silver halides
125	Tetrahedrite-tennantite
100	Polybasite-pearceite and other silver sulfantimonides and sulfarsenides
75	Argentite
58	Chalcopyrite
50	Cerussite
35	Smithsonite
25	Native silver
5	Anglesite

Classified by commodity, the rank, total output from 1858 to 1963, and percent of total of the principal mineral production in Colorado are given in the following table of the U.S. Bureau of Mines.

Petroleum	$1,899,360,000	25	per cent
Coal	1,372,130,000	18	
Gold	918,250,000	12	
Molybdenum	883,000,000	11	
Silver	607,130,000	8	
Zinc	389,280,000	5	
Lead	338,430,000	4	
Sand and gravel	206,780,000	3	
Vanadium	200,240,000	3	
Uranium (1948-1963)	189,450,000	2	
Cement	165,060,000	2	
Natural gas and natural gas liquids	145,900,000	2	
Stone	120,120,000	2	
Copper	101,970,000	1	
Tungsten	49,070,000	1	
Fluorspar	38,120,000	.5	
Miscellaneous (including clay, pyrite, gypsum, perlite)	46,710,000	1	
Total	$7,671,000,000		

Anyone seriously interested in Colorado minerals must own a copy of *Minerals of Colorado; A 100-Year Record,* by Edwin B. Eckel (U. S. Geological Survey Bulletin 1114, 1961). It is a thorough summary of the first century of published knowledge, as well as of considerable unpublished information. It gives the chief occurrence of the common minerals, and all known occurrences of the rest. The references are extensive, numbering 800, and the 32 introductory pages are of great value, especially to the beginner.

A primary source of mineral information are the publications of the U. S. Geological Survey. A monthly catalog will be sent free upon request to the Survey at the Denver Federal Center or Washington 25, D. C.; a 5-year catalog and annual supplements are also available without charge.

Annual reports of Colorado mineral production, but seldom naming the mineral species other than the gems, are issued by the U. S. Bureau of Mines and are available from the Denver Federal Center or 225 Custom House Building, Denver.

Excellent summaries of mineral production, as well as a good deal of geologic information, appear in *Mineral Resources of Colorado* (State of Colorado Mineral Resources Board, 1947) and *First Sequel* (1960).

Individual mineral species are indexed under "Colorado" in the "Bibliographies of North American Geology," published at intervals as Bulletins of the U. S. Geological Survey. Past cumulations, which can be bought or consulted in libraries, carry the literature back to 1785.

The earliest notices of Colorado minerals, pertaining only to native gold, appeared in newspapers and personal correspondence. They were followed by a book by J. P. Whitney, *Silver Mining Regions of Colorado* (Van Nostrand, New York, 1865), in which other minerals were mentioned.

The first list was prepared by J. Alden Smith for Ovanda J. Hollister's fascinating book *The Mines of Colorado,* published in 1867 (Samuel Bowles and Company, Springfield, Mass.).

In 1870, Smith, later to become Colorado's first state geologist, published a similar list, somewhat augmented, in his *Catalogue of the Principal Minerals of Colorado with Annotations on the Local Peculiarities of Several Species.* This booklet of 16 pages, including the paper cover, was "printed at the Register office" in Central City. Between 1870 and 1883, Smith issued several such catalogs, which were cumulated in his list of 1883, in which the names of contributors are given (Ref. 3).

Beginning in 1877, or even earlier, J. S. Randall, publisher of the *Georgetown Courier*, began to print lists and notes pertaining to Colorado minerals. He also issued some paperbound booklets entitled *Minerals of Colorado* and placed announcements in a department, "The Minerals of Colorado," in the *Colorado School of Mines Scientific Quarterly* (Ref. 4).

Members of the Hayden Survey — the U. S. Geological and Geographic Survey of the Territories — prepared periodic lists of minerals found in Colorado Territory, as well as separate lists of those found within the district assigned to the South Park and Middle divisions. These were published in the Annual Reports (Ref. 5).

In 1885, Whitman Cross, a pioneer geologist who lived into the decade of the Second World War, listed (Ref. 6) 55 mineral species that he considered "specially noteworthy" and gave data on their occurrence and references.

The same year, S. F. Emmons described (Ref. 7) the chief minerals of the then-recognized mining counties of Colorado.

Whitman Cross in 1883 (Ref. 8) and 1888 (Ref. 9) listed, by mineral and by county, the known useful minerals being mined or not mined in Colorado at that time. These lists were enlarged upon in 1914 (Ref. 10) and 1917 (Ref. 11) in the lists of useful minerals by Frank C. Schrader, Ralph W. Stone, and Samuel Sanford.

The epochal 6th edition (1892) of *Dana's System of Mineralogy* had an extensive list of Colorado minerals by counties and district.

A detailed guide to collecting localities, featuring many sketch maps and giving information on accessibility, land ownership, and local conditions, is *Colorado Gem Trails and Mineral Guide*, by Richard M. Pearl. It is especially suitable for the amateur and hobbyist.

Of the numerous books on mineralogy, the one that most emphasizes Colorado occurrences is *Minerals and Rocks. Their Nature, Occurrence, and Uses*, by Russell D. George (Appleton-

Century-Crofts, Inc., New York, 1943); he was for many years the state geologist of Colorado. This book is based on his popular Bulletins 6 and 12 of the Colorado Geological Survey.

REFERENCES

Ref. 1. U. S. Geological Survey Bulletin 1114, 1961.
Ref. 2. U. S. Geological Survey Bulletin 1114, 1961, p. 4-6.
Ref. 3. Report on the Development of the Mineral, Metallurgical, Agricultural, Pastoral, and other Resources of Colorado for the Years 1881-1882, Tribune Publishing Company, Denver, 1883, p. 127-149.
Ref. 4. Quarterly of the Colorado School of Mines, vol. 1, 1892, p. 98-106; vol. 2, 1893, p. 117-136.
Ref. 5. U. S. Geological and Geographic Survey of the Territories Third Annual Report, 1869, p. 101-130; First, Second, and Third Annual Reports, 1873, p. 201-228; Sixth Annual Report, 1872 (1873), p. 179-182; Seventh Annual Report, 1873 (1874), p. 267-270, 355-361; Eighth Annual Report, 1874 (1876), p. 178-179; Ninth Annual Report, 1875 (1877), p. 226-235; Tenth Annual Report, 1876 (1878), p. 135-159.
Ref. 6. Colorado Scientific Society Proceedings, vol. 1, 1885, p. 134-144.
Ref. 7. 10th Census of the United States, vol. 13, 1885, p. 64-86.
Ref. 8. Mineral Resources of the U. S., 1882, p. 748-753.
Ref. 9. Mineral Resources of the U. S., 1887, p. 707-714.
Ref. 10. U. S. Geological Survey Bulletin 585, 1914, p. 38-50.
Ref. 11. U. S. Geological Survey Bulletin 624, 1917, p. 81-96.

Chapter 12

Colorado Minerals of Particular Interest

About 20 years ago, the Colorado Mineral Society prepared a traveling exhibit of 20 typical Colorado minerals, to circulate among the member clubs of the Rocky Mountain Federation of Mineralogical Societies. Each of the 20 specimens, selected by Harvey C. Markman, was of high quality, and the box was accompanied by a descriptive list written by the secretary, Richard M. Pearl. It seems worthwhile to repeat the brief remarks here, as an introduction to this chapter. Some of the minerals and comments may seem a bit dated, but they were appropriate at the time and are still of interest.

Making a limited list of this sort is not easy, nor will it satisfy everyone. In the rest of this chapter is given information about enough Colorado minerals to help the student or collector prepare his own list of favorites.

"Colorado is noted for the variety of its mineral deposits. These 20 minerals are selected from a list of hundreds as being representative of the state. Some are worth the attention of collectors for the fine crystals they provide. Some furnish attractive gems. Some are of great economic importance, having made the foundations of boomtowns and the fortunes of mining kings. Some others have helped win wars. All are familiar to Colorado mineral enthusiasts and should be in every collection of Colorado specimens.

1. *Native Gold* — Colorado was born with a gold spoon in its mouth. An example of the $900,000,000 worth that has come from the Colorado Rockies.

2. *Native Silver* — Here is one of the half-billion-dollar reasons why Colorado is called the Silver State. These specimens came from Aspen, where the world's largest silver nugget was found.

3. *Bismuthinite* — Colorado is one of the few localities for this rather rare sulfide of bismuth. It is a heavy mineral and a main source of the metal bismuth.

4. *Molybdenite* — The miners call it "Molly." It resembles graphite but makes a bluish mark on paper. A mountain of it rises above Climax, producing a large share of the world's output from Colorado's biggest mining operation.

5. *Sphalerite* — The most important zinc material. This black, iron variety is called marmatite and comes from Kokomo. The Empire mine at Gilman has the greatest reserves of any zinc mine in the United States.

6. *Pyrite* — Some of the best pyrite crystals ever found have come from mines in Colorado. This specimen is from Central City. This sulfide of iron often contains finely divided gold and other precious substances.

7 *Calaverite* — A telluride of gold with some silver, from Cripple Creek, one of history's truly great mining camps. When "roasted," the tellurium vaporizes, leaving yellow globules of gold.

8. *Sillimanite* — An orthorhombic, aluminum silicate not found in many Western localities. It is used in ceramics and as gems.

9. *Goethite* — The Pikes Peak region has made a common, iron-ore mineral suitable for cabinet specimens, deserving to be in any collection. The color is brown, but the streak is yellow.

10. *Rhodochrosite* — One of Colorado's most sought-after minerals. A managanese carbonate and common enough, but not many localities have produced specimens as good as those from Colorado.

11. *Aragonite* — "Indian dollars" are twinned crystals of aragonite, found along the eastern foothills of the mountains. The peculiar angle of twinning gives them the appearance of a normal hexagonal mineral, but they are an orthorhombic calcium carbonate that has changed to calcite, which is hexagonal.

12. *Amazonstone* — The most characteristic Colorado gem, first introduced at the Centennial Exposition. It is one of the choicest varieties of feldspar, which is ordinarily a common mineral.

13. *Aquamarine* — The leading prize of collectors who find the

73

gem cavities of Mount Antero in the Sawatch Ranch, at an altitude of over 14,000 feet.

14. *Lapis Lazuli* — A Colorado gem, found high on the rugged slopes of North Italian Mountain in Gunnison County. The golden spangles are pyrite, and the white flecks are calcite.

15. *Garnet* — Specimens of garnet from Colorado are widely known for their size, color, and perfection of form. These are the almandite variety from Salida. The characteristic, green coating is an alteration to chlorite.

16. *Phenakite* — A beryllium-bearing, gem mineral from Mount Antero, the highest collecting locality in North America. The first discovery of phenakite in the United States was in Colorado, which is still the leading source.

17. *Analcime* — Representative of the zeolite minerals from the twin Table Mountains of Golden, a world-famous locality for this group. Found in cavities in the volcanic rock.

18. *Uraninite* — A powerfully radioactive mineral from the Central City pitchblende deposit, the largest in the United States. It is a primary source of radium.

19. *Barite* — These blue crystals are typical of Weld County's noted deposit. A surprisingly heavy mineral, it is composed of barium sulfate.

20. *Ferberite* — An iron tungstate from the Boulder County mines that produced most of America's strategic tungsten of the First World War."

Edwin B. Eckel (Ref. 1) called attention to certain of the distinctive features of Colorado's minerals. He remarked upon the presence in abundance of certain minerals that are uncommon or even rare in most of the rest of the world, especially the tellurium minerals. He noted the unusual abundance of sulfosalt minerals, especially those containing bismuth, arsenic, and antimony. The uranium and vanadium minerals of the Colorado Plateau come in for special consideration (see page 104). Colorado pegmatites, which include so many of the gems, are of particular value (see page 109), and other concentrations of interest to Dr. Eckel include the zeolite minerals of the Table Mountains (Jefferson County), the contact-metamorphic minerals of North Italian Mountain (Gunnison County), and the rare minerals of Iron Hill (Gunnison County). Other unusual minerals also occur in Colorado, a number of which are described briefly in this chapter.

Tellurium is the chemical element of perhaps more importance in Colorado than in any other state. Tellurium minerals are abundant here, though rare elsewhere in the country; there are only two other regions in the world — Transylvania (Rumania) and Western Australia — where they occur in significant amounts. Yet, in Colorado, 23 such minerals have been reported, 5 of which were first found here, and 2 of which — calaverite, krennerite — have been responsible for more than $400 million in gold. Most of the krennerite at Cripple Creek, where it was first identified in the United States, was thought for decades to be sylvanite, which was also first identified in the United States in Colorado (but in Boulder County). From a large cavity (called the Jewelry Store) in the Cresson mine came calaverite worth $16,000 per ton. Curiously, the district of Telluride does not contain the minerals for which it was named. Native tellurium is one of the tellurium minerals, especially prominent in Boulder, Gunnison, La Plata, and Montezuma Counties.

The 23 Colorado tellurium minerals are as follows:

Altaite	$PbTe$
Calaverite	$AuTe_2$
Coloradoite	$HgTe$
Colusite	$Cu_3(As,Sn,V,Fe,Te)S_4$
Emmonsite	$Fe_2(TeO_3)_3 \cdot 2H_2O$ (?)
Empressite	$AgTe$
Ferrotellurite	$FeTeO_4$ (?)
Hessite	Ag_2Te
Krennerite	$AuTe_2$
Melonite	$NiTe_2$
Montanite	$(BiO)_2(TeO_4) \cdot 2H_2O$ (?)
Nagyagite	$Pb_5Au(Te,Sb)_4S_{5-8}$
Petzite	Ag_3AuTe_2
Rickardite	Cu_4Te_3
Stuetzite	Ag_5Te_3
Sylvanite	$(Ag,Au)Te_2$
Tellurite	TeO_2
Tellurium	Te
Tellurobismuthite	Bi_2Te_3

Tetradymite	Bi_2Te_2S	Wehrlite	Bi_3Te_2 (?)
Vulcanite	$CuTe$	Weissite	Cu_5Te_3

The sulfosalt minerals, which are so typical of the Mineral Belt, have yielded most of Colorado's silver and copper. Tetrahedrite-tennantite (including freibergite), the so-called gray copper minerals, are the dominant sulfosalts. Tetrahedrite (freibergite) from Georgetown contains up to 30 percent silver. The Mollie Gibson mine, at Aspen, was the first place in the United States to report the existence of tennantite.

Pearceite, which accounts for about $100 million in silver production, was named for a Colorado man, even though it is not a type mineral from this state. It was of particular significance as an ore mineral at Aspen, where it was the chief silver mineral in the Mollie Gibson mine: as much as 16,000 ounces of silver per ton. Pearceite grades into polybasite, which it was formerly called.

The chief sulfosalt minerals in Colorado are the following:

Enargite	Cu_3AsS_4
Pearceite	$(Ag,Cu)_{16}As_2S_{11}$
Polybasite	$(Ag,Cu)_{16}Sb_2S_{11}$
Proustite	Ag_3AsS_3
Pyrargyrite	Ag_3SbS_3
Tennantite	$(Cu,Fe)_{12}As_4S_{13}$
Tetrahedrite	$(Cu,Fe)_{12}Sb_4S_{13}$

The zeolite minerals of North and South Table Mountains are famous. These include analcime (or analcite), chabazite (the most abundant), heulandite, laumontite, levynite, mesolite (specimens have been described as superb), mordenite (the type Colorado mineral ptilolite), natrolite, scolecite, stilbite, and thomsonite. Other minerals of interest are present here, including good crystals of apophyllite. Collecting at this outstanding locality is described in *Colorado Gem Trails and Mineral Guide*. Some of the zeolite minerals — especially analcime, scolecite, and stilbite — also have good occurrences elsewhere in the state.

The metamorphic minerals of North Italian Mountain are diverse but still little known. They include chlorite, diopside, epidote, garnet, graphite, lazurite, scapolite (mizzonite), sahlite (a pyroxene), talc, and idocrase. Of the last mineral, this is said to be one of the finest localities for specimens in the United States. Other interesting minerals occur here, but they are less typically metamorphic. The most valuable product of North Italian Mountain, however, is the fine, blue lapis lazuli, for which this is one of the few localities in the world. This area is far from thoroughly explored; a former president of the Colorado Mineral Society says that this is the only place in North America (from Alaska to Mexico) where he was ever lost!

Iron Hill, also in Gunnison County, contains a rare carbonitite deposit, consisting of an intimate association of carbonate rocks and alkalic igneous rocks. A long list of minerals comes from here. Those not found elsewhere in Colorado are the following: brugnatellite, cancrinite, cebollite, hastingsite, juanite, melilite, perovskite, and synchisite.

Those also found elsewhere in the state include the following: aegirite (a pyroxene), bastnaesite, calcite, cerite (?), diopside (a pyroxene), dolomite, fluorite, garnet, glaucophane (an amphibole), idocrase (titaniferous), monticellite, natrolite (a zeolite), nepheline, olivine, pyrochlore, soda-tremolite (an amphibole), spinel, thorite, thorogummite (?), wollastonite, and xenotime.

The area of St. Peters Dome, in El Paso County, is remarkable for its rare, fluorine-bearing minerals, some new to mineralogy. These include cryolite, elpasolite, gearksutite, pachnolite, prosopite, ralstonite, thomsenolite, and weberite.

The following notes on other Colorado minerals of especial interest do not include those that are discussed in the separate chapters on Colorado "type" minerals (see Chapter 13), radioactive minerals (see Chapter 15), and native metals (see Chapter 17). Some duplication in this chapter is unavoidable. In addition, there are varieties of certain Colorado minerals that are in some places better known than the species themselves — for example, freibergite, the silver-bearing variety of tetrahedrite, and marmatite, the iron-rich variety of sphalerite. Mineral collectors have also made familiar many special variety names of quartz (especially chalcedony) and other minerals.

A major species of amphibole, *actinolite* is widely disseminated in metamorphic rocks in Colorado.

Aegirite is found in a number of Colorado localities, though less than some of the other species of pyroxene.

Remarkably abundant in certain rocks, *allanite* is a familiar sight in numerous places in Colorado.

The second proved occurrence of *alleghanyite* in the United States was in the Sunnyside vein, Silverton district (San Juan County).

Huge amounts of *alunite* occur in Colorado. This mineral was first reported in the United States from the Rosita Hills, Custer County.

Anatase, also called octahedrite, occurs in small amounts in many igneous rocks. The Cebolla district, in Gunnison County, is of especial interest for specimens.

Anglesite has been an abundant mineral in the mines of Colorado, especially in the upper parts mined in earlier years.

A fairly common Colorado mineral, *anthophyllite* is a species of amphibole.

One of the most valuable of Colorado's minerals is *argentite,* found in many ore deposits.

St. Peter's Dome, in El Paso County, is one of the few localities in the world for *astrophyllite..*

A familiar species of pyroxene in Colorado is *augite,* which sometimes is found as good crystals.

Although not abundant or in beautiful specimens, *azurite* is widespread in Colorado.

Barite is an abundant Colorado mineral. The blue crystals from near Stoneham (Weld County) and Hartsel (Park County) rank among the best ever found.

Although not readily recognized, *bertrandite* is an important ore of beryllium at Badger Flats (Park County). Its better known occurrence at Mount Antero (Chaffee County) is the finest locality for the mineral in North America; the heart-shaped, twin crystals are extraordinarily handsome.

Beryl is a gem and an ore mineral of considerable importance in Colorado, especially at Mount Antero (Chaffee County) and Badger Flats (Park County).

A very familiar mineral in Colorado rocks is *biotite,* which sometimes is found in huge masses weighing hundreds of pounds.

Beautiful specimens of *bornite* have been taken from several places in Colorado, notably Gilman (Eagle County) and the Evergreen mine (Gilpin County).

Specimens of *calcite* are widely distributed in Colorado, but most of the calcite exists as limestone and marble rock. Thermoluminescent dog-tooth calcite is found at South Table Mountain (Jefferson County). Phosphorescent and fluorescent calcite is familiar in Larimer County.

Cerargyrite in pieces weighing up to 100 pounds has come from the Leadville district in Lake County. This is one of Colorado's most valuable minerals.

The *cerite* from Jamestown, in Boulder County, was dated at 940 million years by the lead-uranium method.

An abundance of *cerussite*, including some excellent crystals, is a feature of Colorado lead deposits. It played an important part in the renewed development of the Leadville district, in Lake County.

Chalcanthite is found on the walls of copper mines and in the ore deposits at a number of localities.

A widespread mineral in Colorado, *chalcopyrite* is also a very valuable ore, yielding copper, gold, and even platinum.

The *chlorite* alteration on the garnet of the Sedalia mine, near Salida (Chaffee County), is the most interesting occurrence of the mineral in the state.

Among the largest crystals of *chrysoberyl* in the world are those from Drew Hill, in Jefferson County. The discovery of this mineral in Douglas County in 1893 was the first made west of New York.

Although not otherwise an important Colorado mineral, the presence of true sapphire makes *corundum* one of the more interesting.

Large, solid masses of nearly pure *covellite* have come from the Summitville district, in Rio Grande County. The crystals occur in fine blades and rosettes.

Cristobalite is very common in the volcanic rock of the San Juan Mountains but not readily recognized at sight.

Cryolite is the most important of the uncommon, fluorine-bearing minerals at St. Peters Dome, in El Paso County. Much of it, however, has been proved to be weberite.

A widespread mineral in Colorado copper mines is *cuprite*.

Newly found in the oil shale (see page 41), *dawsonite* promises to become one of Colorado's most abundant minerals.

The species of pyroxene called *diopside* is rather abundant in certain Colorado localities.

Extremely abundant as the chief constituent of the rock of the same name, *dolomite* is also found in veins and other ore deposits.

Enargite is rather common in a number of places in Colorado, and in some makes good specimen material.

Widely disseminated, *enstatite* is a fairly common species of pyroxene in Colorado.

Some choice *epidote* occurs in several Colorado localities, especially the Calumet mine, in Chaffee County.

The mineral *epsomite* is natural epsom salts and is common at mineral springs, alkali lakes, and in mines.

The various kinds of *feldspar* are of the utmost importance as Colorado minerals. Not always identified accurately, these include orthoclase, microcline, and plagioclase, and their varieties and subspecies.

More *ferberite* has come from the Nederland tungsten belt of Boulder County than from any other place in the world. The attractive crystals are black and shining.

Fluorapatite is the most common kind of apatite, a widespread mineral in Colorado.

A large amount of *fluorite* is mined in Colorado, where it is one of the more important nonmetallic minerals.

The second known occurrence of *friedelite* in the United States is the Sunnyside mine, in the Silverton district (San Juan County).

Although *gadolinite* is considered a rare mineral, it has been mined commercially in Colorado.

Galena is one of the good specimen minerals in Colorado, as well as one of the most important ores, yielding lead, silver, and gold.

All six species of *garnet* have been found in Colorado. The almandite crystals at the Sedalia mine are world famous, the gemmy spessartite crystals at Nathrop are also well known, and other Colorado occurrences are likewise worthy of attention.

Although *gearksutite* is a rare, fluorine-bearing mineral, it has been found in localities in Colorado.

Pikes Peak (granite) flanked by uptilted sedimentary rocks of the Garden of the Gods. *Stewart's.*

Rhyolite, associated with garnet and other gems, Arkansas Valley, Colorado. *Ward's Natural Science Establishment.*

Graphic granite, the most characteristic aspect of Colorado pegmatite. *Ward's Natural Science Establishment.*

Granite, the most common igneous rock in Colorado. *Ward's Natural Science Establishment.*

Basalt, the porous variety called scoria, Oak Creek, Colorado. *Ward's Natural Science Establishment.*

Shale, the most abundant sedimentary rock in Colorado. *Ward's Natural Science Establishment.*

Sandstone, consisting mostly of quartz, is familiar in Colorado. *Ward's Natural Science Establishment.*

Upturned sedimentary rocks in the foothills at Morrison, Colorado. *T. S. Lovering, U. S. Geological Survey.*

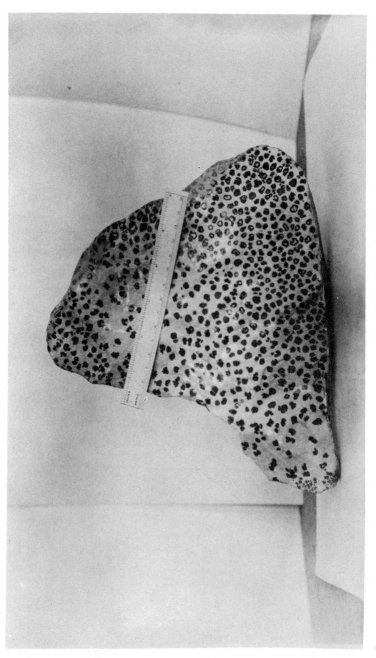

Leopard rock, a curious white sandstone found in Larimer County, Colorado. *E. B. Eckel, U. S. Geological Survey.*

Oil-shale deposit at Rifle, Colorado. *U. S. Bureau of Mines.*

Pumice, a light volcanic rock. *Ward's Natural Science Establishment.*

Gneiss, a typical metamorphic rock in Colorado. *Ward's Natural Science Establishment.*

Some *graphite* has been mined in Colorado in the past, and smaller amounts are seen in various places.

Gypsum is a very abundant mineral in Colorado as the chief constituent of the rock having the same name. As alabaster, selenite, and satin spar, it is specimen material.

Halite is the mineral of which rock salt is composed. It is abundant in Colorado, chiefly in deep beds.

As specimens, masses, and iron ore bodies, *hematite* is one of the most common Colorado minerals.

The *hedenbergite* species of pyroxene is abundant in Colorado but only in the Snowmass Mountain area, in Gunnison County.

Formerly called calamine, *hemimorphite* is a familiar mineral in the zinc deposits of the state.

The most common kind of amphibole, *hornblende* is found widely in Colorado though of little specimen interest.

The distribution of *huebnerite* in the southwestern part of Colorado is substantial, and some of the crystals are large.

Common enough in Colorado rocks, but not conspicuous, is *hypersthene*, a species of pyroxene.

Ice and *water* are minerals of great interest, though usually not thought of as such.

Ilmenite is fairly common in Colorado rocks.

Little noticed, *jarosite* is nevertheless a rather widespread mineral in Colorado.

Not nearly so common as once thought, *kaolinite* is nevertheless probably the most important of the clay minerals in Colorado.

The largest specimens of *lepidolite* ever found — 10 inches across — have come from the Brown Derby mine, in Gunnison County.

Limonite is one of the most common of Colorado minerals but is not specimen material except as a pseudomorph after some other mineral. Most so-called limonite is probably goethite.

Although it imparts a green color to many rocks, *malachite* is not found in substantial quantities in Colorado.

Melanterite is a secondary mineral in various mine workings in Colorado.

Especially at Leadville (Lake County), *minium* is a familiar mineral.

Leading all other Colorado minerals in value of production to date is *molybdenite,* most of which has come from the world's largest deposit, at Climax (Lake and Summit Counties). Not much of it, however, is of specimen quality.

As the chief constituent of many clay beds, *montmorillonite* is abundant in Colorado.

In Colorado, as elsewhere, *muscovite* is a common mineral. It occurs in some pegmatite in huge masses.

Nahcolite is one of the more interesting of Colorado's minerals because of its chemical composition: it is naturally occurring baking soda. It comes from the oil shale of western Colorado, being found in layers as thick as 4 inches and in pieces as large as 5 feet across.

Opal (hydrophane) "from some Colorado locality" was described in 1887 by George F. Kunz and in 1889 by A. H. Church, as a curious mineral that absorbs water, turning from chalky white to entirely clear. Kunz said, "The finder has proposed the name 'magic stone' for it and has suggested its use in rings, lockets, and charms to conceal photographs, hair, or other objects which the wearer wishes to reveal only when his caprice dictates." This occurrence is not now known.

The extremely rare mineral *pachnolite* of other parts of the world is the third most abundant mineral in certain pegmatites at St. Peters Dome (El Paso County).

The first discovery of *phenakite* in the United States was in 1882 at Crystal Park (El Paso County). It has since become recognized as an abundant mineral at Crystal Peak (Teller County) and Mount Antero (Chaffee County).

Phosgenite is common at Ilse (Custer County) but is not know elsewhere in the state.

Pickeringite was first reported in the United States in 1877 from specimens collected near the Garden of the Gods, at Colorado Springs.

St. Peters Dome, El Paso County, is one of the few places in the world where unaltered specimens of *prosopite* have been found.

Colorado *pyrite* has been studied carefully, for it includes some of the largest and finest crystals of this mineral ever found — some as much as 5.5 inches across. It is also valuable for its gold, copper, and silver, and its use for fluxing ores and as a source of sulfuric acid.

Fairly widespread, particularly at Leadville (Lake County), *pyromorphite* is an important Colorado mineral.

The best known of all minerals, in Colorado as well as elsewhere, *quartz* is found in all types of rock. The chalcedony varieties are, like the crystalline varieties, important gems and make fine specimens.

Ralstonite is a rare, fluorine-bearing mineral from St. Peters Dome, in El Paso County.

The choicest *rhodochrosite* ever found has come from the Sweet Home mine, near Alma (Park County). This is also an abundant mineral elsewhere in the state.

Rhodonite is a fairly common mineral, especially in the Silverton district, in San Juan County. Its possibilities as a gem are good.

Rare in most of the world, *riebeckite* — a species of amphibole — is abundant near Colorado Springs. It was once thought to be arfvedsonite.

An interesting mineral common in Colorado is *roscoelite*, a vanadium-bearing mica.

Recognized usually by its fluorescence, *scheelite* has often been the clue to the presence of other tungsten minerals with which it occurs.

Serpentine is a major mineral in many parts of the world, but its Colorado occurrences have not been identified with much certainty.

Seldom identified by its correct name, *siderite* is found in concretions, surrounded by limonite. These are especially common near Pueblo.

Occurring in many zinc deposits in Colorado, *smithsonite* is one of the most valuable minerals in the state.

At Cripple Creek (Teller County), especially, *sodalite* is an abundant mineral in volcanic rocks.

Masses of *stibnite* weighing up to 50 pounds have been found in the mines of Cripple Creek.

Stromeyerite has been an important ore of silver in Colorado but unfortunately not common enough.

Sulfur (native) is widely distributed in Colorado, especially around hot springs.

The second occurrence in the world of *thalenite* is near Woodland Park, in Teller County.

Another of the rare, fluorine-bearing minerals from St. Peters Dome is *thomsenolite*.

The largest crystals of gem *topaz* ever found in North America have come from Colorado pegmatite, and this mineral has other types of occurrences here as well.

A leading gem mineral, *tourmaline* is common in Colorado. Most of the specimens are black.

Tremolite, a species of amphibole, is abundant in Colorado only in the Monarch district, in Chaffee County.

Although *tridymite* is usually considered a rare mineral, it is so abundant in the volcanic rocks of the San Juan Mountains that an estimated 350 cubic miles of it are believed to be contained in them.

Perhaps the first mineral used by man in Colorado was *turquoise.* It remains one of the most highly prized.

Weberite includes much of what was formerly thought to be cryolite at St. Peter's Dome, in El Paso County.

The third known locality for *willemite* in the United States was the Sedalia copper mine, near Salida (Chaffee County).

Zinkenite was first authenticated in the United States from the Brobdignag claim, in the Red Mountain district (San Juan County).

The frequency with which it is met makes *zircon* worthy of mention as a Colorado mineral.

REFERENCES

Ref. 1.   U. S. Geological Survey Bulletin 1114, 1961, p. 12-14.

Chapter 13

Type Minerals of Colorado

Beginning in 1877 with coloradoite, 45 new minerals from this state have become type minerals: those first found in and described from Colorado. An additional 20 minerals, presumed to be type minerals, have been discredited. Most of these 65 minerals bear the names of Colorado men and places, histories of which have been given elsewhere by Richard M. Pearl (Ref. 1-4).

More significant, actually, are the type minerals. Of these, pearceite and carnotite are of outstanding economic value. The minerals named after Colorado places are, of course, also type minerals, but not all those named after Colorado men are — for some minerals, such as pearceite, were discovered elsewhere. In addition, there are certain Colorado type minerals that are not based on personal or geographic names -- such as, for example, cuprobismutite. These 65 minerals of especial Colorado interest are described below, with the present status of each indicated.

Alaskaite — named in 1881 for the Alaska mine, in Pough-keepsie Gulch, in the Silverton district — is probably a mixture of several metallic minerals. The original discovery, which has since been proved to be such a mixture, yielded as much as 3,000 ounces of silver per ton of ore. Similar material has come from Bolivia, and the name has been given to specimens from two other Colorado mines.

Beegerite — named in 1881 after Hermann Beeger, a Denver metallurgist, who donated it to the University of Pennsylvania

93

— is a lead-bismuth sulfosalt. It was first found in the Baltic lode, which is on Revenue Mountain, near the head of Geneva Gulch, Montezuma district, Park County. Two other Colorado localities are known, apart from which the only occurrences seem to be in Switzerland and Siberia.

Beidellite — named in 1925 for a now-vanishing mining camp in the San Juan Mountains, Saguache County — is an aluminum-bearing variety of montmorillonite, which is a clay mineral of considerable importance. However, the desirability of usinig this name is in dispute.

Berryite—named in 1966 after L. G. Berry, professor of mineralogy at Queen's University, in Canada, who obtained the first X-ray powder pattern of it—is a lead-copper-silver-bismuth sulfosalt. This mineral was found while making a new study of cuprobismutite from the Missouri mine, Park County, which proved it to be a valid species. In turn, berryite displaced galenobismutite as an acceptable name for impure material found earlier in Sweden.

Brockite—named in 1962 after Maurice Brock, of the U. S. Geological Survey—is a calcium-thorium phosphate. It was found in a prospect pit near the Bassick mine, Querida, Custer County. Similar material may have been seen elsewhere earlier.

Carnotite — named in 1889 after Marie-Adolpe Carnot, a French mining engineer — is a major uranium mineral, a hydrous potassium-uranium vanadate. The type locality was somewhere in Montrose County, and Pierre and Marie Curie determined its radioactivity from the original specimens.

Cebollite — named in 1914 for Cebolla Creek, Gunnison County — is a hydrous calcium-aluminum silicate. Its only known occurrence is Iron Hill, a locality of especial interest.

Coffinite — named in 1954 after Reuben Clare Coffin, 'a geologist and a member of a distinguished northern-Colorado family — is a hydrous uranium silicate. Not discovered until the uranium boom, it was soon shown to be an abundant mineral in many ores. The type locality was the La Sal No. 2 mine, at the head of Lumsden Canyon, near Gateway, Mesa County.

Coloradoite — named in 1877 for the newly created state — is a mercury telluride. It was first reported in the Keystone and Mountain Lion mines, in the Magnolia district, and the Smuggler mine, in the Ballarat district, both in Boulder County. A rare mineral here and in one or two other counties, it has been found to be rather abundant in the La Plata district of La Plata and Montezuma Counties.

Corvusite — named in 1933 because if its purplish-bluish-black color (like that of a raven) — is a hydrous vanadium vanadate. It was described from material found in Gypsum Valley, San Miguel County, and in Utah. A good deal of it has been shown to exist in western Colorado.

Creedite — named in 1916 for Creede, in Mineral County — is a hydrous sulfate of complex composition. Its type locality is the fluorite-barite mine at Wagon Wheel Gap. Here, and in the few other places it is known in the world, it is very rare.

Cuprobismutite — named in 1863 because of its chemical composition — is a copper-bismuth sulfosalt, but its validity has been questioned. The type locality is the Missouri mine, Hall Valley, Park County.

Danalite — named in 1866 after Prof. James D. Dana, one of America's outstanding geologists and mineralogists — is an iron-beryllium silicate of the helvite group. It was described from Stone Mountain, El Paso County, but the original material was later shown to be genthelvite, and so the name danalite was given to a different member of the same group.

Delrioite — named in 1959 after A. M. del Rio, a pioneer American mineralogist — is a hydrous calcium-strontium vanadate. It was discovered on the dump of the Jo Dandy mine, on the southwest wall of Paradox Valley, in Montrose County.

Doloresite — named in 1957 for the Dolores River — is a hydrous vanadate. Its type locality is the La Sal No. 2 mine, near Gateway, Mesa County (the original home of coffinite), and it was soon found elsewhere.

Domingite — named in 1889 for the Domingo mine, on the ridge between Dark Canyon and Baxter Basin, the Ruby-Elk Mountain district, Gunnison County —is now discredited as a new mineral. It has been confused with a number of other minerals; the extremely complicated mix-up, which produced two wrong names (warrenite being the other) and authenticated occurrences for three minerals, has been outlined by Dr. Eckel (page 81).

Doughtyite — named in 1905 for the owner of the Doughty Hot Springs, Delta County — is a hydrous aluminum sulfate. It seems to be the same as winebergite, so that either name may

95

become the correct one for this somewhat uncertain mineral.

Duttonite — named in 1956 after Clarence Edward Dutton, a leading American geologist — is a vanadium hydroxide. It was found in the Peanut mine, Montrose County.

Elpasolite — named in 1883 for El Paso County — is a potassium-sodium-aluminum fluoride. It was discovered on St. Peters Dome and has never been seen anywhere else.

Empressite — named in 1914 for the Empress Josephine mine, Bonanza district, Saguache County — is a silver telluride. It has also been found in another mine in the same county and in Boulder County.

Ferrotellurite — named in 1877 because of its chemical composition (iron and tellurium) — is a discredited name for a supposedly new mineral found in the Keystone mine, Magnolia district, Boulder (also the source of coloradoite).

Fervanite — named in 1961 because of its chemical composition — is a hydrous iron vanadate. Its original discovery was in the Tiny mine, Gypsum Valley, San Miguel County, and it has been found elsewhere in the Colorado Plateau.

Fremontite — named in 1916 for Fremont County — is an obsolete name for natromontebrasite, which was also wrongly called natroamblygonite. The type locality was Eight Mile Park, near Canon City.

Genthelvite — named in 1944 after Frederick A. Genth, a well-known mineral chemist — is a zinc-beryllium silicate of the helvite group. It was originally called danalite, a name that has since been applied to another member of the same group. The type locality is Stove Mountain, in El Paso County; two other occurrences have been found close by.

Gilpinite — named in 1917 for Gilpin County — was shown in 1926 to be johannite. The exact locality of the type material is unknown.

Goldschmidtite — named in 1899 after Victor Goldschmidt, a leading crystallographer — was shown the next year to be the same as sylvanite. Its type locality was the Gold Dollar mine, Arequa Gulch, Cripple Creek district, Teller County.

Guitermanite — named in 1884 after Franklin Guiterman, metallurgist and chemist at Silverton and Denver, who brought it to light — is of uncertain identity. It came from the Zuñi mine, on Anvil Mountain, near Silverton, San Juan County.

Gunnisonite — named in 1882 for Gunnison County — has not been satisfactory sustained as a distinct mineral. The type locality was Iron Hill, in Gunnison County, also the source of cebollite.

Hendersonite—named in 1962 after Edward P. Henderson, of the U. S. National Museum, who had long studied vanadium minerals—is a calcium vanadyl vanadate. It was first found in 1955 at the J. J. mine, on the south side of Paradox Valley, Montrose County, and later in New Mexico.

Henryite — named in 1874 for Prof. Joseph Henry, secretary of the Smithsonian Institution — was thought to be a new iron-lead telluride but was shown the same year to be a mixture of two other minerals. The original locality was the Red Cloud mine, Gold Hill district, Boulder County.

Hinsdalite — named in 1911 for Hinsdale County — is a hydrous phosphate and sulfate. It was first found at the Golden Fleece mine, Lake City district, where it is abundant; yet, it is not known for certain any place else.

Hummerite — named in 1950 for the Hummer mine, of the Jo Dandy group, Bull Canyon district, Montrose County (also the source of delrioite) — is a hydrous potassium-magnesium vanadate. It was later found elsewhere in the county.

Ilesite — named in 1881 after Dr. Malvern W. Iles, metallurgist of the Grant Smelting Company, Leadville, who analyzed it and described it in print — is a hydrous sulfate of manganese, zinc, and iron. The type locality is on the McDonnell mining claim, near Middle Swan Creek, Hall Valley, Montezuma district, Park County, where it is abundant.

Juanite — named in 1934 for the San Juan Mountains — is a hydrous aluminum silicate. It came from Iron Hill, in Gunnison County (also the source of cebollite and gunnisonite), where it is common.

Lillianite — named in 1889 for the Lillian mine, on Printer Boy Hill, Leadville district, Lake County — is of uncertain identity. The original material was shown to be a mixture of three minerals, but something like lillianite occurs elsewhere as a lead-bismuth sulfosalt.

Lionite — named in 1877 for the Mountain Lion mine, Magnolia district, Boulder County, also an original source of coloradoite — is probably a mixture of minerals.

Magnolite — named in 1877 for the Magnolia district, Boulder

County — is not a distinct mineral. The locality was the Keystone mine, an original source of coloradoite.

Metahewettite — named in 1914 because of its chemical and physical relationship to hewettite — is a hydrous calcium vanadate. It was first identified in Paradox Valley, Montrose County. It seems a widespread mineral in the Colorado Plateau, but much so-called metahewettite is actually barnesite.

Metarossite — named in 1927 because of its chemical and physical relationship to rossite — is a hydrous calcium vanadate. The type locality is Bull Pen Canyon, San Miguel County, and the mineral has been found at another place in the county.

Metatyuyamunite — named in 1956 because of its chemical and physical relationship to tyuyamunite — is a hydrous calcium uranate and vanadate. Its first discovery was at the Jo Dandy mine, Montrose County (the source of delrioite and hummerite), and it has since been found in many places on the Colorado Plateau, as well as in Teller County.

Montroseite — named in 1950 for Montrose County — is a vanadium hydroxide. Its type locality was the Bitter Creek mine in Paradox Valley. Other mines of the Colorado Plateau in Colorado and Utah have also yielded it.

Natramblygonite — named in 1911 because of its chemical relationship to amblygonite — was renamed fremontite in 1916. Even this name was rejected in 1955 in favor of natromontebrasite. The type locality is Eight Mile Park, near Canon City, in Fremont County.

Nicholsonite — named in 1913 after U. S. Senator Samuel D. Nicholson, of the Western Mining Company, who brought it to attention — is a zinc-bearing variety of aragonite. Discovered in the Leadville district, Lake County, it has also been described from Garfield County. Curiously, some of the original material contains strontium rather than zinc.

Paramontroseite — named in 1955 because of its chemical and physical relationship to montroseite — is a vanadium oxide. Discovered in the Bitter Creek mine, Paradox Valley, Montrose County, where montroseite also came from, it has been found elsewhere in Montrose County and in Mesa County.

Ptilolite — named in 1886 because if its light, downy nature — was proved in 1958 to be mordenite. The type locality was North and South Table Mountains and Green Mountain, near

Golden, Jefferson County. More of the same material comes from the Silver Cliff district, in Custer County.

Rickardite — named in 1903 after T. A. Rickard, former state geologist of Colorado, who brought it to attention — is a copper telluride. It was found in the Good Hope mine at Vulcan, Gunnison County (also the type locality of weissite and vulcanite), and has come also from Saguache County.

Rilandite — named in 1933 for the owner of the J. L. Riland mining claim, near Meeker, Rio Blanco County, where it was found — is of uncertain status. Some has also been found in Routt County but nowhere else.

Robellazite — named in 1900 after a French engineer who studied the occurrence — is evidently a mixture. Its type locality was somewhere in Montrose County, but the vaguely described material has not been accepted as a distinct mineral.

Rossite — named in 1927 after Clarence S. Ross, a distinguished geologist — is a hydrous calcium vanadate. The original and only locality is in the Bull Pen Canyon area, San Miguel County, where it appears to be fairly common.

Sanfordite — named early in the 1900's after Albert B. Sanford, a Denver assayer, who was involved in making it known — is another name for rickardite. Both names are used because of the conflict as to priority in making the identification.

Sauconite — named in 1946 for Saucon Valley, Pennsylvania — is a zinc-bearing variety of montmorillonite, a clay mineral. The original descriptions were based on specimens in several mines of the Leadville district, Lake County.

Schirmerite — named in 1874 after Prof. J. F. L. Schirmer, superintendent of the U. S. Mint at Denver, who brought it to attention — is a lead-silver-bismuth sulfosalt. An earlier attempt to give this name to a mixture of minerals from the Red Cloud mine in Boulder County met with failure. True schirmerite came first from the Treasure Vault lode, in what is now the Montezuma district, in Park and Summit Counties. Other occurrences in Colorado are less certain.

Sherwoodite — named in 1958 after Alexander M. Sherwood, a mineral chemist — is a hydrous calcium vanadate. Although first found in Mesa County, the type material came from the Peanut mine, in Montrose County, and later finds have been made in Colorado, Utah, and New Mexico.

Siderophyllite — named in 1880 because of its geologic occurrence ("iron loving") — is a high iron-bearing variety of biotite mica. It was first described from Pikes Peak, in El Paso County.

Simplotite — named in 1958 after J. R. Simplot, owner of the second mine in which it was found — is a hydrous calcium vanadate. The type locality is the Sundown claim, San Miguel County, and it has since been found in several places in Montrose County.

Steigerite — named in 1935 after George Steiger, a mineral chemist — is a hydrous aluminum vanadate. Found first on the Sullivan Brothers' claims, Gypsum Valley, San Miguel County, it has since been seen nearby and in Montrose County.

Telaspyrine — named in 1877 because of its supposed chemical composition (tellurium, arsenic, sulfur, iron) — is regarded as a mixture of several minerals and has no standing. The type locality was the Sunshine district, Boulder County.

Tysonite — named in 1880 after S. T. Tyson, its discoverer — is a rare-earth fluoride. It came probably from Stove Mountain, in El Paso County, also the source of danalite and genthelvite. The approved name for this mineral is generally fluocerite.

Vandiestite — named in 1899 after Peter H. Van Diest, director of the Lead Mining Company, of San Luis — was shown in 1940 to be a mixture of several minerals. Also recorded as VonDiestite, it came originally from the Hamilton and Little Gerald mines, on Sierra Blanca, where Huerfano, Costilla, and Alamosa Counties come together.

Vanoxite — named in 1924 because of its chemical composition — is a hydrous vanadium vanadate. The type locality is the Jo Dandy mine, Paradox Valley, Montrose County. Two other localities are also known, besides this original source of delrioite, hummerite, and metatyuyamunite.

Vulcanite — named in 1961 for Vulcan, Gunnison County — is a copper telluride. It came from the Good Hope mine, which was also the type locality of rickardite and weissite.

Warrenite — named in 1890 after E. R. Warren, of Crested Butte, who sent samples of it to the U. S. Geological Survey, in Denver — shares in the confusion associated with domingite and is likewise discredited as a distinct mineral.

Weissite — named in 1927 after Dr. Loui Weiss, owner of the Good Hope mine — is a copper telluride. Its type locality is the

Good Hope and Mammoth Chimney mines, at Vulcan, Gunnison County, where also rickardite was discovered earlier and vulcanite later.

Wolftonite — named in 1913 for the Wolftone mine, Leadville district, Lake County — was found in 1942 to be hydrohetaerolite. It is abundant but known only at Leadville.

Zunyite — named in 1884 for the Zuñi mine on Anvil Mountain near Silverton, San Juan County, which was also the original home of guitermanite — is a basic silicate of aluminum. It has since been found nearby and in Ouray and Saguache Counties.

REFERENCES

Ref. 1.   Colorado Magazine, vol. 18, 1941, p. 48-53, 137-142.
Ref. 2.   Mineralogist, vol. 16, 1948, p. 59-61.
Ref. 3.   Mineralogist, vol. 19, 1951, p. 283-286.
Ref. 4.   Mineralogist, vol. 20, 1952, p. 70-72.

Chapter 14

Rare-Earth Minerals

The rare-earth minerals represent an interesting group of Colorado minerals and one that promises to be economically important in future years. Composed of the 14 to 17 so-called rare-earth elements, these minerals yield interesting products in the atomic-energy program, in metallurgy, and for other uses.

Vance Haynes, Jr. (Ref. 1) has enumerated the following 25 named rare-earth minerals known in Colorado.

Allanite	$(Ca,Ce,Th)_2(Al,Fe,Mg)_3Si_3O_{12}(OH)$
Bastnaesite	$(Ce,La)FCO_3$
Brannerite	$(U,Ca,Fe,Y,Th)_3(Ti,Si)_5O_{16}$ (?)
Cerite	$CeSiO_4$ (?)
Crytolite	$ZrSiO_4$
Doverite	$CaYF(CO_3)_2$
Euxenite	$(Y,Ca,Ce,U,Th)(Nb,Ta,Ti)_2O_6$
Fergusonite	$(Y,Er,Ce,Fe)(Nb,Ta,Ti)O_4$
Fluocerite	$(Ce,La,Nd)F_3$
Gadolinite	$Y_2FeBe_2O(SiO_4)_2$
Microlite	$(Na,Ca)_2Ta_2O_6(O,OH,F)$
Monazite	$(Ce,La,Th)PO_4$
Pyrochlore	$(Na,Ca)_2(Nb,Ta)_2O_6F$
Samarskite	$(Y,Ce,U,Ca,Fe,Pb,Th)(Nb,Ta,Ti,Sn)_2O_6$
Samiresite	$(U,Ca)(Nb(Ta,Ti)_3O_9 \cdot nH_2O$
Tengerite	$CaY_3(CO_3)_4(OH)_3 \cdot 3H_2O$ (?)
Thorite	$ThSiO_4$
Thorogummite	$Th(SiO_4)_{1-x}(OH)_{4 \cdot x}$
Törnebohmite	$(Ce,La,Al)_3(OH)(SiO_4)_3$

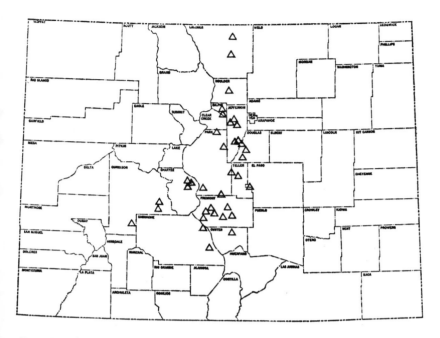

Rare-Earth Deposits in Colorado. *Vance Haynes, Jr., State Mineral Resources Board.*

Uranothorite	$(Th,U)SiO_4$
Xenotime	YPO_4
Yttrocerite	$(Ca,Ce,Y)F_2$
Yttrofluorite	$(Ct,Y,Ce)F_2$
Yttrotantalite	$(Fe,Y,U)(Nb,Ta)O_4$
Yttrotitanite	$(Ca,Y,Ce)(Ti,Al,Fe)O(SiO_4)$

REFERENCES

Ref. 1.   Mineral Resources of Colorado. First Sequel, State Mineral Resources Board, Denver, 1960, p. 383.

103

Chapter 15

Radioactive Minerals

Minerals containing uranium or thorium are radioactive. So also are those containing certain other elements, but they are not of present commercial interest. Colorado has been one of the leading producers of uranium from a variety of minerals, and it may be a major source of thorium in the future.

The first discovery of pitchblende in the United States was at the Wood mine, Central City, in 1871. The crystallized mineral, of purer composition, is called uraninite; large cubes are a feature of a deposit in Routt County. Carnotite is a Colorado "type mineral," having been described in 1899 from Roc Creek (Montrose County) specimens, and carnotite from western Colorado was the world's principal source of radium during the early years of that metal's use in medicine. Mined also for uranium and vanadium, sometimes as a byproduct, carnotite took on a new aspect with the development of nuclear energy in 1945. With the uranium boom of the 1950's, a host of other radioactive minerals, many of them new ones, came to attention.

The main uranium districts are the following, of which those in the Uravan mineral belt have yielded most of the uranium

Uravan mineral belt
    Gateway district: Mesa County
    Uravan district: Montrose County
    Paradox district: Montrose County
    Bull Canyon district: Montrose, San Miguel Counties

Gypsum Valley district: San Miguel County
Slick Rock district: San Miguel County
San Juan Mountains
    Rico district: Dolores, San Juan Counties
    Lightner Creek district: La Plata County
Meeker district: Rio Blanco County
Maybell district: Moffat County
Cochetopa district: Saguache County
Marshall Pass district: Chaffee, Gunnison, Saguache Counties
Front Range district: Boulder, Clear Creek, Gilpin, Jefferson,
    Larimer Counties
Tallahassee district: Fremont County
The main thorium districts are the following:
    Wet Mountains: Custer, Fremont Counties
    Powderhorn district: Gunnison County
    St. Peters Dome: El Paso County

The principal radioactive minerals in these deposits are the following; many others, including some very familiar to mineral collectors, are radioactive but not yet commercial.

Autunite	$Ca(UO_2)_2(PO_4)_2 \cdot 10\text{-}12H_2O$
Bequerelite	$7UO_3 \cdot 11H_2O$
Carnotite	$K_2(UO_2)_2(VO_4)_2 \cdot 1\text{-}3H_2O$
Coffinite	$U(SiO_4)_{1-x}(OH)_{4x}$
Euxenite	$(Y,Ca,Ce,U,Th)(Nb,Ta,Ti)_2O_6$
Johannite	$Cu(UO_2)_2(SO_4)_2(OH)_2 \cdot 6H_2O$
Meta-autunite	$Ca(UO_2)_2(PO_4)_2 \cdot 2\frac{1}{2}\text{-}6\frac{1}{2}H_2O$
Metatorbernite	$Cu(UO_2)_2(PO_4)_2 \cdot 4\text{-}8H_2O$
Pitchblende	$UO_2 \cdot UO_3$
Rauvite	$CaO \cdot 2UO_2 \cdot 5V_2O_5 \cdot 16H_2O$
Schroeckingerite	$NaCa_3(UO_2)(CO_3)_3(SO_4)F \cdot 10H_2O$
Thorianite	$(Th,U)O_2$
Thorite	$ThSiO_4$
Thorogummite	$ThSiO_4 \cdot nH_2O$
Torbernite	$Cu(UO_2)_2(PO_4)_2 \cdot 12H_2O$
Uraninite	$UO_2 \cdot UO_3$
Uranophane	$Ca(UO_2)_2(SiO_3)_2(OH)_25H_2O$
Uranopilite	$(UO_2)_6(SO_4)(OH)_{10} \cdot 12H_2O$
Volborthite	$CU_3(VO_4)_2 \cdot 3H_2O$
Zippeite	$3UO_3 \cdot 2SO_3 \cdot 9H_2O$ (?)

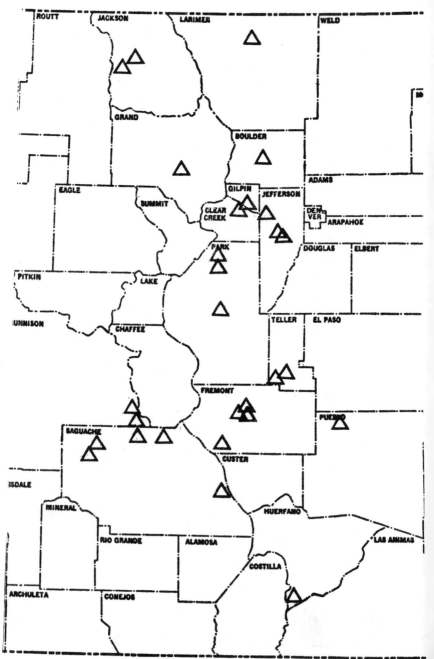

Uranium Deposits in Central Colorado. *Robert J. Wright and Donald L. Everhart, State Mineral Resources Board.*

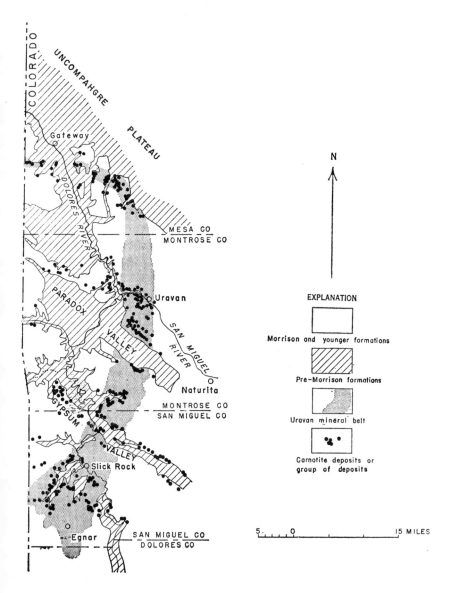

Uranium Deposits in Western Colorado. *U. S. Geological Survey.*

A good summary of Colorado uranium deposits is Chapter 5, by Robert J. Wright and Donald L. Everhart, in *Mineral Resources of Colorado. First Sequel* (State Mineral Resources Board, Denver, 1960).

Practically all Colorado occurrences are mentioned in *Mineralogy and Geology of Radioactive Raw Materials*, by E. William Heinrich (McGraw-Hill Book Company, New York, 1958).

A glossary of radioactive minerals is U. S. Geological Survey Bulletin 1250 (1967), by Judith W. Frondel, Michael Fleischer, and Robert S. Jones.

Chapter 16

Gem Minerals

Colorado is one of the best sources of gems for the collector. In tabulations published by the U. S. Geological Survey (Ref. 1) and the U. S. Bureau of Mines (Ref. 2), Colorado ranked as one of the seven states having the greatest variety of such occurrences.

Probably the most characteristic Colorado gem is amazonstone. This decorative, green variety of microcline feldspar comes from cavities, or pockets, in the Pikes Peak granite. In the centennial year of 1876, Pikes Peak amazonstone made its first public appearance at the exposition in Philadelphia. The abundant supply and superior quality of the Colorado crystals drove from the market a large stock of Russian specimens brought all the way from the Urals. Crystal Peak, Devils Head, and Crystal Park are among the places where splendid specimens of this "Pikes Peak jade" or "Pikes Peak turquoise" can be secured.

The 14,269-foot summit of Mount Antero is the highest gem locality in North America and second highest in the world. Its rich, blue crystals of aquamarine have not been available to collectors in recent years because of beryl-mining operations, but this was long an outstanding place in which to find gems.

Turquoise is an important Colorado gem, mostly in the central zone of the state. The King mine, near Manassa, is said to be the deepest turquoise mine in existence; from here, in 1946, came the largest nugget of this gem ever seen up to that time. The Hall mine, near Villagrove, yields turquoise of unexcelled quality.

The delicate blue color and exquisite transparency of Colorado topaz has made possible the cutting of many lovely gems. Perhaps the largest well-formed and well-preserved topaz crystal ever found in North America was taken from a cavity in the weathered granite of Devils Head.

Surely a most unusual gem material is agatized dinosaur bone. Once a part of the huge beasts that lumbered through the swamps that marked the landscape before the rise of the present generation of Rocky Mountains, these bones are readily recognized by their peculiar cell structure, perfectly preserved in the swirling patterns of banded agate. Localities for this gem are mentioned in Part IV of this book.

Among the other Colorado gems of especial interest are its lapis lazuli from North Italian Mountain, Gunnison County (see page 76); its sapphire from Turret, Chaffee County; its jet from the Pikes Peak region; its alabaster from Owl Canyon (Larimer County) and elsewhere; its garnet from Nathrop, Chaffee County; its amethyst from Redfeather Lakes, Larimer County; and its widespread rock crystal, smoky quartz, and rose quartz, as well as agate, jasper, petrified wood, and other chalcedony. These and many others are described in *Colorado Gem Trails and Mineral Guide,* by Richard M. Pearl (Sage Books, Denver, 2d edition, 1964), a detailed guide to the more profitable and accessible localities. Information on turquoise in Colorado and other areas is given in *Turquoise and the Indian,* by Edna Mae Bennett (Sage Books, Denver, 1966).

REFERENCES

Ref. 1.   U. S. Geological Survey Bulletin 1042-G, 1957.
Ref. 2.   U. S. Bureau of Mines Bulletin 585, 1960, p. 327.

Chapter 17

Native Metals

No class of minerals is of greater interest to collectors than the native metals, the first minerals known to have been used by early man. Most of these occur in Colorado. Several of the more important are among Colorado's most distinctive minerals.

Native gold, some of remarkable purity, is widespread in Colorado. Historically, it is the most significant of the hundreds of minerals in the state. Large nuggets are not typical of this gold, but crystals of superlative quality are familiar from Clear Creek, Boulder, La Plata, and Montezuma Counties, and especially from Farncomb Hill, in the Breckenridge district of Summit County. Magnificent specimens from here are to be seen in the Denver Musum of Natural History and the Colorado School of Mines.

What has been reported as the largest nugget of native silver ever found is the 1,870-pound mass recovered from the Smuggler mine at Aspen in 1894. Native silver is widespread in Colorado, not only as masses of the nature, though not of this size, but as wire and leaf silver of exceptional beauty. The best localities have been in Boulder, Clear Creek, Gunnison, Lake, Mineral, Pitkin, and Summit Counties.

The occurrence of native copper in Colorado has more mineralogic interest than commercial value. Some of the pieces have weighed as much as 500 pounds. In certain mines, the iron rails have been replaced by copper precipitating out of the acid minewater.

The only occurrences of native iron in Colorado are in the

meteorites arriving here from outer space. These are enumerated in Chapter 9.

Platinum and the other five minerals of the platinum group — iridium, palladium, osmium, rhodium, ruthenium — have been reported in Colorado on occasion, but authenticated occurrences are few. A certain amount of fraud has been suspected in some of the announcements.

A curiosity, because it is a liquid mineral, is native mercury. This occurs in the tellurium districts of Colorado, probably originating by alteration of other mercury-bearing minerals.

Natural alloys of mercury and silver, known as amalgam, have been found in Boulder, La Plata, and Montrose Counties. Some Colorado amalgam also contains gold.

Publicity has been given to the alleged finding of native lead in Colorado, but it cannot now be verified.

Chapter 18

Important Colorado Metals

GOLD

Historically and economically the most important metal in Colorado, gold has been mined here since 1858. This state is second to California, and ahead of Alaska, in total production. It has 46 of the 505 principal districts in the country; Cripple Creek ranks second in production (19,088,462 fine ounces through 1958), Southern Gilpin is 11th (4,240,037 ounces), Telluride is 14th (3,738,768 ounces), Leadville is 16th (2,939,968 ounces), and Sneffels is 25th (1,899,146 ounces).

Gold occurs by itself, with silver, and with base metals. The gold-bearing minerals include native gold, the tellurides, and certain base-metal sulfides. The most valuable of thes minerals in Colorado are the following:

Native gold	Au
Calaverite	$AuTe_2$
Krennerite	$AuTe_2$
Sylvanite	$(Ag,Au)Te_2$
Pyrite	FeS_2
Chalcopyrite	$CuFeS_2$

SILVER

Colorado, the Silver State, ranks third among the states in the production of this metal, which is second to gold in importance.

113

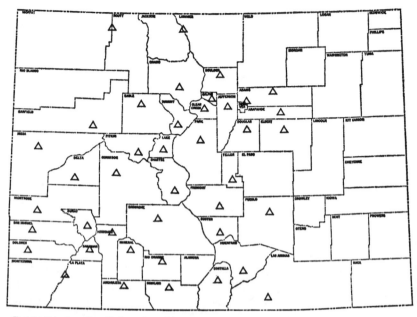

Gold districts in Colorado. *U. S. Geological Survey.*

Silver has been mined here since 1859. Colorado has 7 of the 25 principal districts in the nation; Leadville stands 5th (239,890,-000 fine ounces), Aspen is 9th (101,304,000 ounces), Red Cliff is 11th (61,220,000 ounces), Creede is 14th (57,860,000 ounces), Telluride is 15th (56,300,000), Silver Plume-Georgetown is 24th (31,000,000 ounces), Idaho Springs-Lamartine is 25th (25,000,-000 ounces). Hence, Lake County has been the main source, although San Miguel and Eagle Counties have led since the Second World War.

Silver is today more than a precious metal for jewelry and coins. It forms the base for the photographic industry, and it is an important metal in the chemical industry, including medicinal uses. Dentistry, solder, and electrical goods help the demand for this versatile metal.

Silver is today more than a precious metal for jewelry and coins. It forms the base for the photographic industry, and it is an important metal in the chemical industry, including medicinal uses. Dentistry, solder, and electrical goods help the demand for this versatile metal.

Silver Districts in Colorado. *Colorado School of Mines.*

The silver-bearing minerals include native silver, the tellurides, other silver compounds, and base-metal minerals.

The main silver-bearing minerals in Colorado are the following:

Native silver	Ag
Sylvanite	$(Ag,Au)Te_2$
Cerargyrite	$AgCl$
Argentite	Ag_2S

Polybasite	$(Ag,Cu)_{16}Sb_2S_{11}$
Pearceite	$(Ag,Cu)_{16}As_2S_{11}$
Galena	PbS
Tetrahedrite	$(CuFe)_{12}Sb_4S_{13}$
Tennantite	$(Cu,Fe)_{12}As_4S_{13}$

COPPER

Copper is sixth in total value among the metals of Colorado, although this state has not been one of its principal producers, ranking only tenth in total output to date.

The standard for electrical conductivity, copper appears as wire and in other forms. Its alloys — brass, bronze, and many more — are used in coinage, jewelry, clocks, and other articles almost too numerous to mention.

Three groups of copper-bearing ores are the main economic ones: native copper, sulfides-sulfosalts, and oxidized compounds. The chief Colorado minerals are the following:

Native copper	Cu
Chalcopyrite	$CuFeS_2$
Bornite	Cu_5FeS_4
Chalcocite	Cu_2S
Covellite	CuS
Tetrahedrite	$(Cu,Fe)_{12}Sb_4S_{13}$
Tennantite	$(Cu,Fe)_{12}As_4S_{13}$
Enargite	Cu_3AsS_4
Chalcanthite	$CuSO_4 \cdot 5H_2O$
Chrysocolla	$CuSiO_3 \cdot 2H_2O$
Cuprite	Cu_2O
Azurite	$Cu_3(CO_3)_2(OH)_2$
Malachite	$Cu_2(CO_3)(OH)_2$
Tenorite	CuO

LEAD

Lead ranks fifth among the metals in Colorado in terms of total production, and Colorado has been fourth among the states in recent years.

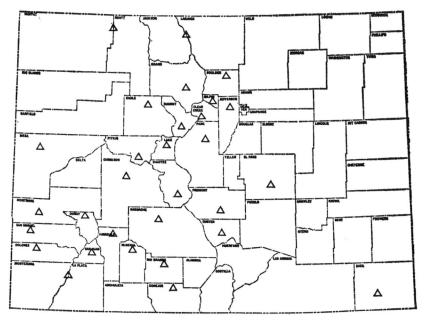

Copper districts in Colorado. *U. S. Geological Survey.*

The ores of lead are usually associated with those of zinc, sometimes with those of copper and other metals. Silver is likewise a valuable constituent of lead deposits in Colorado. Lead was first mined in 1869. Lake County, of which Leadville is county seat, has been by far the chief producer, followed by Pitkin, San Juan, and San Miguel Counties.

The use of lead is well known in batteries and ethyl gasoline, but it has numerous other industrial applications.

The main Colorado lead minerals are these:

Galena PbS
Cerussite $PbCO_3$
Anglesite $PbSO_4$
Pyromorphite $Pb_5Cl(PO_4)_3$

ZINC

Zinc stands fourth among Colorado metals in total value to date, and Colorado is one of the leading states in total produc-

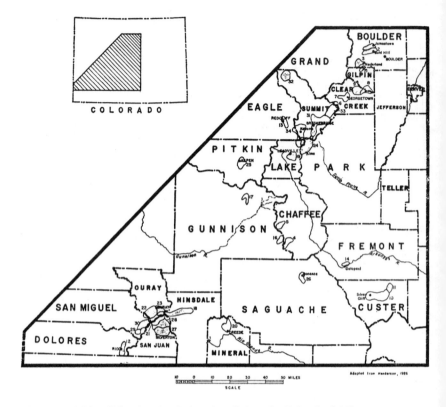

Lead and Zinc Districts in Colorado. *Colorado School of Mines.*

tion, being fourth in recent years. Lead is a common associate. Zinc mining began in Colorado in 1885. Lake and Eagle Counties are responsible for much the largest part of the output, and 2 of the 25 largest mines in the country are in Colorado.

Not a familiar metal in everyday use, zinc nevertheless plays a diverse role in industry. It protects steel, when galvanized, from rusting. Alloyed with copper, it becomes brass. It makes die castings; and it has many other uses too.

The chief zinc minerals are the following:

Sphalerite	ZnS
Smithsonite	$ZnCO_3$
Hemimorphite	$Zn_4(Si_2O_7)(OH)_2 \cdot H_2O$

118

The most valuable single mineral in Colorado's mining history is molybdenite. This fact seems more remarkable when one realizes that Climax has been its sole source and that the deposit is low grade. In consequence, Climax has became the largest single mining operation in the state, it is the largest underground mine in the country, and it is the largest producer of molybdenum in the world. Furthermore, it is given credit for having made possible, by its very size, the molybdenum industry as we know it today. A new deposit of vast size has been opened at Urad.

The Climax mine dates from 1917, when exploration began and was soon followed by production. Activity ceased from 1919 to 1924; the use of molybdenum as an alloy in steel in automobiles then began a new era for the metal.

Associated with molybdenite at Climax are minor amounts of tungsten and tin.

The following molybdenum-bearing minerals occur at Climax:

Molybdenite	MoS_2
Ferrimolybdite	$Fe_2(MoO_4)_3 \cdot 8H_2O$

REFERENCES

Gold, Ref. 1; Silver, Ref. 2; Lead, Ref. 3; Zinc, Ref. 4; Molybdenum, Ref. 5-7.

Ref. 1. Colorado School of Mines Mineral Industries Bulletin, vol. 4, no. 3, 1961.

Ref. 2. Colorado School of Mines Mineral Industries Bulletin, vol. 5, no. 5, 1962.

Ref. 3. Colorado School of Mines Mineral Industries Bulletin, vol. 6, no. 6, 1963.

Ref. 4. Colorado School of Mines Mineral Industries Bulletin, vol. 6, no. 6, 1963.

Ref. 5. Colorado School of Mines Mineral Industries Bulletin, vol. 7, no. 1, 1964.

Ref. 6. Mineral Resources of Colorado. First Sequel, State Mineral Resources Board, Denver, 1960, p. 317-325.

Ref. 7. Quarterly of the Colorado School of Mines, vol. 37, no. 4, 1942.

Chapter 19

Colorado Geologic Museums

Boulder: Henderson Museum
      Location: University of Colorado Campus
      Hours: 8 - 5, Monday-Saturday; 2 - 5, Sunday
      Exhibits: 3,600 rocks, minerals, and gems; fossils
      Admission: Free

*Canon
City:*   Municipal Museum
      Location: Sixth and River Streets
      Hours: 8 - 12, Monday-Saturday; 1 - 5 every day
      Exhibits: Rocks, minerals, fine dinosaur fossils
      Admission: Free

*Colorado
Springs:*   Pioneers' Museum
      Location: 25 West Kiowa Street
      Hours: 10 - 5, Tuesday-Saturday; 3 - 6, Sunday
      Exhibits: Approximately 150 specimens in mineral
          and gem collection; emphasis on Pikes Peak
          region crystals
      Admission: Free

Denver: Denver Museum of Natural History
      Location: City Park
      Hours: 10-4:30, Monday-Saturday; 12-5, Sundays
          and holidays
      Exhibits: 7,250 rocks, minerals, and gems; outstand-
          ing fossils
      Admission: Free

During the early years of Denver, mineral collections owned by the State Historical and Natural History Society were on display successively in the Glenarm Hotel (at that time the state capitol), the Arapahoe Court House, the Denver Chamber of Commerce, and the State Capitol Building. Later transferred to the State Museum Building, this superb collection has now been moved to the Colorado School of Mines. Emphasizing an arrangement by counties, it numbers about 12,000 specimens, posing a problem in display that has not yet been solved.

Fort
Collins: Fort Collins Pioneer Museum
          Location: 219 Peterson Street
          Hours: 1-5, every day; 7-9 during summer
          Exhibits: Minerals related to mining
          Admission: Free

Golden: Colorado School of Mines Geologic Museum
          Location: Berthoud Hall, 16th and Maple Streets
          Hours: 8-5, Monday-Friday; 1:30-4, Sat and Sun
          Exhibits: Rock, mineral, and gem collection; approximately 5,200 specimens; fossils
          Admission: Free

Gunnison: Western State College Museum
          Location: College Campus
          Hours: 2 - 5, Monday-Friday, during school
          Admission: Free

Saguache: Saguache County Museum
          Hours: 10 - 12, 1 - 5, everyday, June - August; open evenings, Saturday and Sunday
          Admission: Free

Steamboat
Springs: Tread of Pioneers Museum
          Location: 507 Pine Street
          Hours: 4 - 9, May 30 - Labor Day
          Admission: Free

Sterling: Overland Trail Museum
          Location: 1.5 miles east of Sterling
          Hours: 10 - 4, Tuesday-Sunday, May 1-October 31
          Exhibits: Rocks, minerals, fossils. Largest piece of agatized wood in the state (12 tons).
          Admission: Free

Chapter 20

Tables for Identifying Colorado Minerals

The following simple tables should enable you to identify most Colorado mineral specimens, or at least to place them within a narrow group for a more detailed examination. Without elaborate equipment and considerable experience, no one can expect to identify all the minerals he is apt to find, but these 100-plus, carefully selected minerals represent nearly all the kinds that are common or abundant. The author is indebted to Edwin B. Eckel for his help in choosing the most appropriate minerals. Others of especial interest are described in Chapter 12.

The terms that are used to describe the tests are defined below. Minerals that are apt to appear in more than one group are repeated where necessary. If the test seems a borderline one, try both groups. Within each group, the minerals are further listed according to the familiar Dana classification. Because of its structure, quartz (together with the related tridymite, cristobalite, and opal) is placed with the silicates instead of the oxides. More extensive tables are given in books devoted to mineralogy, but none of these books includes all the minerals given here, although most of the minerals will be found. Keys to many of the minerals, together with drawings that show the points of identification to look for, will be found in *How to Know the Minerals and Rocks*, by Richard M. Pearl (McGraw-Hill Book Company; New American Library of World Literature, Inc.)

Elastic: snaps back after bending.

Flexible: bends without being elastic

Luster: surface reflection of light; the color is not considered.

Metallic luster: resembling the surface of a metal; these minerals are either opaque or else give a dark streak.

Nonmetallic luster: does not resemble the surface of a metal; these minerals give a light or white streak or no streak at all.

Radioactive: capable of being recorded on a Geiger counter or similar instrument.

Resinous luster: oily to greasy surface.

Scratch: the test for hardness; do not mistake a mark (like chalk on a blackboard) for a scratch; use a fresh surface and avoid crumbly material.

Streak: color of the powdered mineral; it is best seen by rubbing the mineral on unglazed porcelain, such as a rough tile or broken china dish.

Submetallic luster: somewhat resembling the surface of a metal.

Tarnish: chemical alteration on surface; it may change the color.

Luster: Metallic or submetallic

Group I: Cannot scratch copper cent

Native elements

	White Colorless	Gray	Black	Red Pink	Brown Orange	Yellow	Green	Blue	Purple Violet
Native gold — Au						x			
Native silver — Ag	x	x	x						
Native copper — Cu				x	x	x			
Native tellurium — Te	x	x							

Sulfides - tellurides

	White Colorless	Gray	Black	Red Pink	Brown Orange	Yellow	Green	Blue	Purple Violet
Argentite — Ag_2S		x	x						
Petzite — $(Ag,Au)_2Te$		x							
Chalcocite — Cu_2S		x	x						
Galena — PbS		x							
Covellite — CuS								x	x
Bismuthinite — Bi_2S_3		x							
Molybdenite — MoS_2		x							
Krennerite — $AuTe_2$	x	x				x			
Calaverite — $AuTe_2$	x	x				x			
Sylvanite — $(Ag,Au)Te_2$	x	x							

Sulfosalts

	White Colorless	Gray	Black	Red Pink	Brown Orange	Yellow	Green	Blue	Purple Violet
Polybasite — $(AgCu)_{16}Sb_2S_{11}$			x						
Pyrargyrite — Ag_3SbS_3		x		x					
Proustite — Ag_3AsS_3		x		x					

Special Features

Heavy; can be hammered; golden-yellow streak
Sharp; can be hammered; tarnish likely; gray or white streak
Can be hammered; tarnish likely; copper-red streak
Gray streak

Can be cut with a knife; tarnish likely; dark-gray streak
Tarnish likely; gray streak
Can be cut with a knife; tarnish likely; dark-grey or black streak
Often breaks into blocks; dark-gray or black streak
Gray or black streak
Can be cut with a knife; gray streak
Marks paper; greenish-gray streak
Often fine, parallel lines; gray streak
Often fine, parallel lines; yellowish to greenish-gray streak
White to yellowish to gray streak

Black streak
Often in bands; red streak
Red streak

Oxides - hydroxides

	White Colorless	Gray	Black	Red Pink	Brown Orange	Yellow	Green	Blue	Purple Violet
Hematite Fe_2O_3		x	x	x	x				
Wad — Mn oxides		x	x					x	
Goethite-Limonite $HFeO_2$					x	x			
Montroseite $VO(OH)$			x						

Silicates

Biotite
$K(Mg,Fe)_3(AlSi_3O_{10})(OH)_2$

Chlorite
$Mg_3(Si_4O_{10})(OH)_2Mg_3(OH)_6$

	White Colorless	Gray	Black	Red Pink	Brown Orange	Yellow	Green	Blue	Purple Violet
Biotite			x		x		x		
Chlorite		x	x				x		

Red to brown streak

Variable mixture of manganese minerals

Yellowish-brown streak

Black streak

Elastic flakes or sheets; gray or white streak

Flexible flakes or sheets; pale-green or white streak

		White Colorless	Gray	Black	Red Pink	Brown Orange	Yellow	Green	Blue	Purple Violet
Native elements										
Native gold	Au						x			
Native silver	Ag	x	x	x		x	x			
Native copper	Cu			x	x				x	x
Sulfides - tellurides										
Petzite	$(Ag,Au)_2Te$		x							
Chalcocite	Cu_2S		x	x						
Sphalerite	ZnS			x		x	x	x		
Chalcopyrite	$CuFeS_2$						x			
Pyrrhotite	FeS					x	x			
Arsenopyrite	$FeAsS$	x	x				x			
Krennerite	$AuTe_2$	x	x				x			
Sulfosalts										
Polybasite	$(AgCu)_{16}Sb_2S_{11}$			x						
Pearceite	$(Ag,Cu)_{16}As_2S_{11}$			x						
Tetrahedrite	$(Cu,Fe)_{12}Sb_4S_{13}$		x	x						
Tennantite	$(Cu,Fe)_{12}As_4S_{13}$		x	x						
Enargite	Cu_3AsS_4			x						

Special Features

Heavy; can be hammered; golden-yellow streak
Sharp; can be hammered; tarnish likely; gray or white streak
Can be hammered; tarnish likely; copper-red streak

Gray streak
Tarnish likely; dark-gray or black streak
Resinous luster; white or gray to yellow to brown streak
Tarnish likely; greenish-black streak
Weakly magnetic; tarnish likely; gray or black streak
Tarnish likely; gray or black streak
Often fine, parallel lines; gray streak

Black streak
Black streak
Brown to gray or black streak
Brown to gray or black streak
Gray or black streak

Oxides - hydroxides

	White Colorless	Gray	Black	Red Pink	Brown Orange	Yellow	Green	Blue	Purple Violet
Hematite Fe_2O_3		x	x	x	x				
Ilmenite $FeTiO_3$			x		x				
Wad — Mn oxides	x	x						x	
Uraninite UO_2			x		x		x		
Goethite-Limonite $HFeO_2$			x		x	x			
Magnetite Fe_3O_4			x						
Euxenite $(Y,Ca,Ce,U,Th)(Nb,Ta,Ti)_2O_6$			x		x				

Carbonates

	White Colorless	Gray	Black	Red Pink	Brown Orange	Yellow	Green	Blue	Purple Violet
Siderite $FeCO_3$	x				x				

Tungstates

	White Colorless	Gray	Black	Red Pink	Brown Orange	Yellow	Green	Blue	Purple Violet
Huebnerite $MnWO_4$			x	x	x	x	x		
Wolframite $(Fe,Mn)WO_4$			x		x				
Ferberite $FeWO_4$			x						

Silicates

	White Colorless	Gray	Black	Red Pink	Brown Orange	Yellow	Green	Blue	Purple Violet
Sphene $CaTiO(SiO_4)$					x	x	x		
Allanite $X_2Y_3O(SiO_4)(Si_2O_7(OH)$	x	x	x		x		x		
Pyroxene — silicate			x		x		x		
Amphibole — silicate			x		x		x		
Biotite $K(Mg,Fe)_3(AlSi_3O_{10})(OH)_2$			x		x		x		

Special Features

Red to brown streak

Often weakly magnetic; brown to black streak

Variable mixture of manganese minerals

Radioactive; green to brown to black streak

Yellowish-brown streak

Magnetic; black streak

Brownish streak

Fizzes in acid; white or gray to yellow to brown streak

Greenish-gray or yellowish-brown streak

Reddish-brown to black streak

Dark-brown streak

White or gray streak

Often yellow to brown coating; often radioactive; gray to brown
streak

White or gray to green streak

White or gray to green to yellow to brown streak

Elastic flakes or sheets; gray or white streak

Group 3: Cannot be scratched by knife blade

		White Colorless	Gray	Black	Red Pink	Brown Orange	Yellow	Green	Blue	Purple Violet
Sulfides										
Pyrite	FeS_2						x			
Marcasite	FeS_2		x				x			
Oxides - hydroxides										
Hematite	Fe_2O_3	x	x	x						
Ilmenite	$FeTiO_3$			x		x				
Rutile	TiO_2			x	x	x				
Wad	Mn oxides	x	x						x	
Uraninite	UO_2			x		x				
Magnetite	Fe_3O_4			x						
Columbite	$(Fe,Mn)Nb_2O_6$			x		x				
Euxenite $(Y,Ca,Ce,U,Th)(Nb,Ta,Ti)_2O_6$				x		x				
Silicates										
Garnet	$A_3B_2(SiO_4)_3$			x		x				
Allanite $X_2Y_3O(SiO_4)(Si_2O_7)(OH)$		x	x	x	x		x		
Tourmaline $XY_3Al_6(BO_3)_3(Si_6O_{18})(OH)_4$			x		x			x	
Pyroxene	silicate			x		x		x		
Amphibole	silicate			x		x		x		

Special Features

Fool's gold; tarnish likely; greenish- or brownish-black streak
Fool's gold; tarnish likely; greenish- or brownish-black streak

Red to brown streak
Often weakly magnetic; brown to black streak
White or gray to yellow to brown streak
Variable mixture of manganese minerals
Radioactive; brown to black streak
Magnetic; black streak
Brown to black streak
Brownish streak

White streak
Often has yellow to brown coating; gray to brown streak

Often triangular; white or gray streak

White or gray to green streak
White or gray to green to yellow to brown streak

Luster: Nonmetallic

Group 4: Cannot scratch copper cent

	White Colorless	Gray	Black	Red Pink	Brown Orange	Yellow	Green	Blue	Purple Violet
Sulfides									
Covellite CuS								x	x
Sulfosalts									
Pyrargyrite Ag_3SbS_3		x		x					
Proustite Ag_3AsS_3		x		x					
Oxides - hydroxides									
Hematite Fe_2O_3				x	x				
Wad Mn oxides		x	x					x	
Corvusite $V_2V_{12}O_{34} \cdot nH_2O$ (?)	x				x			x	
Goethite-Limonite $HFeO_2$					x	x			
Halides									
Cerargyrite $AgCl$		x			x	x			

Special Features

Gray or black streak

Red streak
Red streak

Red to brown streak
Variable mixture of manganese minerals
Blue-black to brown streak
Yellowish-brown streak

Cuts like wax; white or gray streak

Sulfates

Gypsum $CaSO_4 \cdot 2H_2O$

Phosphates - vanadates

Metatorbernite
$Cu(UO_2)_2(PO_4)_2 \cdot 4\text{-}8H_2O$

Carnotite
$K_2(UO_2)_2(VO_4)_2 \cdot 1\text{-}3H_2O$

Tyuyamunite
$Ca(UO_2)_2(VO_4)_2 \cdot 7\text{-}10.5H_2O$

Metatyuyamunite
$Ca(UO_2)_2(VO_4)_2 \cdot 5\text{-}7H_2O$

Pascoite
$Ca_2V_6O_{17} \cdot 11H_2O$

Metahewettite
$CaV_6O_{16} \cdot 9H_2O$

Silicates

Amphibole   silicate

Clay   silicate

Muscovite
$KAl_2(AlSi_3O_{10})(OH)_2$

Biotite
$K(Mg,Fe)_3(AlSi_3O_{10})(OH)_2$

Lepidolite
$K_2Li_3Al_3(AlSi_3O_{10})_2(OH,F)_4$

Roscoelite
$K_2V_4Al(Si_6O_{20})(OH)_4$

Chlorite
$Mg_3(Si_4O_{10})(OH)_2Mg_3(OH)_6$

Opal $SiO_2 \cdot nH_2O$

	White Colorless	Gray	Black	Red Pink	Brown Orange	Yellow	Green	Blue	Purple Violet
Gypsum	x	x		x					
Metatorbernite							x		
Carnotite						x	x		
Tyuyamunite						x	x		
Metatyuyamunite						x	x		
Pascoite				x	x	x			
Metahewettite				x					
Amphibole	x				x	x	x		
Clay	x	x			x	x			
Muscovite	x	x		x			x		
Biotite			x		x		x		
Lepidolite	x			x					x
Roscoelite			x			x	x		
Chlorite		x	x				x		
Opal	x		x		x	x	x		

White streak

Fine flakes; radioactive; green streak

Powdery masses; radioactive; yellow streak

Powdery masses; radioactive; yellow streak

Powdery masses; radioactive; yellow streak

Crusts; yellow streak

Powdery masses; brownish-red streak

White or gray to green to yellow to brown streak
Slippery feel; clay odor; white or pale-yellow streak
Elastic flakes or sheets; white streak

Elastic flakes or sheets; gray or white streak

Fine flakes; white streak

Fine scales; white streak

Flexible flakes or sheets; pale-green or white streak

Waxy luster; white streak

	White Colorless	Gray	Black	Red Pink	Brown Orange	Yellow	Green	Blue	Purple Violet
Sulfides									
Sphalerite ZnS			x		x	x	x		
Oxides - hydroxides									
Hematite Fe_2O_3				x	x				
Wad Mn oxides		x	x					x	
Corvusite $V_2V_{12}O_{34} \cdot nH_2O$ (?)			x		x			x	
Uraninite UO_2			x		x				
Goethite-Limonite $HFeO_2$			x		x	x			
Euxenite $(Y,Ca,Ce,U,Th)(Nb,Ta,Ti)_2O_6$			x		x				
Halides									
Fluorite CaF_2	x			x	x	x	x	x	x
Carbonates									
Calcite $CaCO_3$	x		x	x		x		x	
Siderite $FeCO_3$		x			x				
Rhodocrosite $MnCO_3$				x					
Smithsonite $ZnCO_3$	x	x			x		x	x	
Cerussite $PbCO_3$	x	x				x			
Dolomite $CaMg(CO_3)_2$	x	x	x		x		x	x	
Ankerite $CaCO_3 \cdot (Mg,Fe,Mn)CO_3$	x	x	x	x	x	x		x	
Bismutite $BiCO_3$		x			x		x		

138

Special Features

Resinous luster; white or gray to yellow to brown streak

Red to brown streak
Variable mixture of manganese minerals
Blue-black to brown streak

Radioactive; brown to black streak
Yellowish-brown streak
Brownish streak

White streak

Fizzes in acid; white streak
Fizzes in acid; white or gray to yellow to brown streak
Fizzes in acid; white streak
Fizzes in acid; white or gray streak
Fizzes in acid; white or gray streak
Fizzes in acid; white or gray streak
Fizzes in acid; white streak

Fizzes in acid; gray streak

Sulfates

Mineral	Formula	White / Colorless	Gray	Black	Red / Pink	Brown / Orange	Yellow	Green	Blue	Purple / Violet
Barite	$BaSO_4$	x	x		x	x	x		x	
Anglesite	$PbSO_4$	x	x							
Alunite	$KAl_3(OH)_6(SO_4)_2$	x			x					

Phosphates - vanadates

| Monazite | $(Ce,La,Y,Th)PO_4$ | | | | x | x | | | | |
| Fluorapatite | $Ca_5F(PO_4)_3$ | x | x | | x | x | x | x | x | x |

Tungstates

Huebnerite	$MnWO_4$			x	x	x	x	x		
Wolframite	$(Fe,Mn)WO_4$			x		x				
Ferberite	$FeWO_4$			x						
Scheelite	$CaWO_4$	x			x	x				

Special Features

White streak
White streak
White streak

White streak

White streak

Greenish-gray or yellowish-brown streak
Reddish-brown to black streak
Dark-brown streak
Glows in ultraviolet light; white streak

Silicates

	White Colorless	Gray	Black	Red Pink	Brown Orange	Yellow	Green	Blue	Purple Violet
Sphene $CaTiO(SiO_4)$					x	x	x		
Cerite $(Ca,Fe)Ce_3H(OH)_2(SiO_4)(Si_2O_7)$ (?)		x		x	x				x
Hemimorphite $Zn_4(Si_2O_7)(OH)_2 \cdot H_2O$	x	x	x						
Allanite $X_2Y_3O(SiO_4)(Si_2O_7)(OH)$		x	x	x	x		x		
Pyroxene silicate	x	x	x		x	x	x		
Rhodonite $Mn(SiO_3)$				x					
Wollastonite $Ca(SiO_3)$	x								
Amphibole silicate	x	x	x		x	x	x		
Muscovite $KAl_2(AlSi_3O_{10})(OH)_2$	x	x		x			x		
Biotite $K(Mg,Fe)_3(AlSi_3O_{10})(OH)_2$			x		x		x		
Lepidolite $K_2Li_3Al_3(AlSi_3O_{10})_2(OH,F)_4$	x			x					x
Roscoelite $K_2V_4Al(Si_6O_{20})(OH)_4$			x			x	x		
Opal $SiO_2 \cdot nH_2O$	x		x		x	x	x		
Analcime $Na(AlSi_2O_6)H_2O$	x	x			x				
Scolecite $Ca(Al_2Si_3O_{10}) \cdot 3H_2O$	x								
Chabazite $(Ca,Na)_2(AlSi_4O_{12}) \cdot 6H_2O$	x			x					
Stilbite $Ca(Al_2Si_7O_{18}) \cdot 7H_2O$	x			x			x		
Coffinite $U(SiO_4)_{1-x}(OH)_{4x}$			x		x				

White or gray streak
White streak

White streak

Often yellow to brown coating; often radioactive; gray to brown
    streak
White or gray to green streak
Often black veins; white streak
White streak
White or gray to green to yellow to brown streak
Elastic flakes or sheets; white streak

Elastic flakes or sheets; gray or white streak

Fine flakes; white streak

Fine scales; white streak

Waxy luster; white streak
White streak

White streak

Nearly cubes; white streak

Sheaf-like form; white streak

Radioactive

Group 6: Cannot be scratched by knife blade

		White Colorless	Grey Black	Red Pink	Brown Orange	Yellow	Green	Blue	Purple Violet
Oxides - hydroxides									
Hematite	Fe_2O_3			x	x				
Rutile	TiO_2		x	x	x	x			
Wad	Mn oxides	x	x					x	
Uraninite	UO_2		x		x				
Goethite-Limonite	$HFeO_2$		x		x				
Chrysoberyl	$BeAl_2O_4$					x	x		
Euxenite $(Y,Ca,Ce,U,Th)(Nb,Ta,Ti)_2O_6$			x		x				
Phosphates									
Turquoise $CuAl_6(PO_4)_4(OH)_8 \cdot 4H_2O$							x	x	

Special Features

Red to brown streak
White or gray to yellow to brown streak
Variable mixture of manganese minerals
Radioactive; brown to black streak
Yellowish-brown streak
White streak
Brownish streak

White or pale-green streak

Silicates

	White Colorless	Gray	Black	Red Pink	Brown Orange	Yellow	Green	Blue	Purple Violet
Phenakite Be_2SiO_4	x	x				x			
Olivine $(Mg,Fe)_2SiO_4$						x	x		
Garnet $A_3B_2(SiO_4)_3$	x		x	x	x	x	x		
Zircon $ZrSiO_4$			x	x	x	x	x		
Silimanite Al_2SiO_5	x						x		
Topaz $Al_2(SiO_4)(F,OH)_2$	x	x		x	x	x		x	
Epidote $Ca_2(Al,Fe)Al_2O(SiO_4)Si_2O_7(OH)$			x			x	x		
Allanite $X_2Y_3O(SiO_4)(Si_2O_7)(OH)$		x	x	x	x		x		
Beryl $Be_3Al_2(Si_6O_{18})$	x	x		x	x	x	x	x	
Cordierite $MgAl_3(AlSi_5O_{18})$		x						x	x
Tourmaline $XY_3Al_6(BO_3)_3(Si_6O_{18})(OH)_4$	x		x	x			x	x	
Pyroxene silicate	x	x	x		x	x	x		
Rhodonite $Mn(SiO_3)$				x					
Amphibole silicate	x	x	x		x		x		
Quartz, chalcedony SiO_2	x	x	x	x	x	x	x	x	x
Tridymite SiO_2	x								
Cristobalite SiO_2	x								
Opal $SiO_2 \cdot nH_2O$	x		x		x	x	x		
Feldspar silicate	x	x		x	x	x	x	x	
Coffinite $U(SiO_4)_{1-x}(OH)_{4x}$			x		x				

Special Features

White streak

White or pale-yellow streak

White streak

Usually pyramids; white streak

Often compact fibers; white streak

White streak

White or gray streak

Often yellow to brown coating; often radioactive; gray to brown streak

White streak

White streak

Often triangular; white or gray streak

White or gray to green streak

Often black veins; white streak

White or gray to green to yellow to brown streak

White streak

Very small; white streak

Very small; white streak

White streak

White streak

Radioactive

Part IV

FOSSILS

Chapter 21

Fossils and Where They Are Found

Fossils have already been stated to be the remains or indications of ancient life. We might explore this simple definition a bit further by asking a few questions:

How old must such life be? Date in years is not an essential factor. A fossil should, of course, be accepted as a geologic specimen rather than belonging to biology, but this is sometimes a matter of interpretation. Fossils give us valuable information about the past, although not until about 1800 were they generally recognized as evidence of former life.

Must such life be extinct? No; many fossils represent still-existing species of plants or animals, whereas others have entirely vanished from the earth.

How are the remains preserved? By rapid burial, serving to protect them from weathering, bacteria, and scavengers. Such burials may preserve the entire body, as has happened to the frozen carcasses of mammoths and mastodons in Siberia. Usually, however, only the hard parts, such as shells, bones, and teeth, are preserved. Suitable parts may be *petrified* by the filling in or the replacement of organic substance by mineral matter; petrified wood is a familiar example.

Most fossils are buried in marine sediments; much of this may take place in the vast seaways that, from time to time, flood the land and divide the continents into irregular islands. Such seaways have been a marked feature of the geologic history of Colorado, as is discussed in the section on rocks. Other fossils may be preserved in streams, swamps, and lakes, in sand dunes, in volcanic ash (as are the Florissant fish and butterflies), in tar, coal, or amber, and in other geologic environments and materials. Petrification is almost always the result of underground, mineral-laden water percolating through soil and rock, depositing some of its chemicals — which are most often silica (SiO_2) — where conditions are favorable.

What *indications* are accepted as fossils? Footprints, trails, and borings are common fossils, as are the carbonized-film imprints of leaves and certain animals. Eggs, impressions of dinosaur and other skins, and coprolites (fossil dung) are still other examples of fossils as indications of life. The actual organisms are not present, but there is no doubt that they once existed. Some geologists refer to "fossil raindrops," "fossil soils," and similar inorganic features of rocks, but these are not true fossils, and such terms, though obvious in meaning, are better omitted.

Fossils serve several useful purposes to the student of earth science. When the plants and animals were living, they tended to be in adjustment to their environment; hence, fossils tell us about the vanished climates, life zones, and related aspects of times long gone. Fossils therefore tell us about the changing geography of our planet, such as the distribution of lands and seas. Fossils are the most significant proof of the principle of organic evolution. By means of fossils, the ordering and relative dating of rocks is mainly accomplished.

As you read the geologic time scale (see page 16), the fossil record does not become satisfactory until the Cambrian Period. Prior to that time — although life had been in existence at least 1 billion years, and perhaps twice as long — the Precambrian rocks register little more than the presence of primitive algae and fungi, and the burrows or trails of worms. Then, at the beginning of the Cambrian, the curtain seems suddenly to have lifted upon a complex and fascinating array of organisms whose fossils lie abundant in the rocks. It seems likely that Precambrian organisms lacked the hard parts necessary for preservation, although other factors may also have contributed to the scantiness of the early record.

Fossils also are useful economically as builders of sedimentary rocks (such as limestone, which may turn to marble), as a means of concentrating various mineral deposits (such as uranium), and as gem materials (such as agatized wood).

Fossils are confined almost solely to sediments and sedimentary rock. The areas marked igneous or volcanic on the geologic map of Colorado (see page 21) can be disregarded. Good places to look include quarries and coal mines, canyons and smaller stream valleys, road cuts, and building excavations.

Locality information of the greatest possible use to the collector of fossils appears in the Folios of the Geologic Atlas of the United States, published by the U. S. Geological Survey. The Colorado folios are listed below; these are all out of print but may be consulted in libraries:

No.	7	Pikes Peak	1894	
	8	Anthracite-Crested Butte	1894	
	36	Pueblo	1897	
	48	Tenmile District Special	1898	
	57	Telluride	1899	
	58	Elmoro	1899	
	60	La Plata	1899	(1901)
	68	Walsenburg	1900	
	71	Spanish Peaks	1901	
	120	Silverton	1905	
	130	Rico	1905	
	131	Needle Mountains	1905	
	135	Nepesta	1906	
	153	Ouray	1907	
	171	Engineer Mountain	1910	
	186	Apishapa	1912	
	198	Castle Rock	1915	
	203	Colorado Springs	1916	
	214	Raton-Brilliant-Koehler, N.M.-Colo. (narrow edge of Colorado)	1922	

The Folios of the Geologic Atlas (listed above) have been superseded by the series called Geologic Quadrangle Maps, of which the Colorado ones are listed below. The descriptions are briefer. Most of them cost $1.00 each and are sold by the U. S. Geological Survey.

GQ	33	Bull Canyon	1954
	55	Gateway	1955
	57	Atkinson Creek	1955

58	Red Canyon	1955	
59	Gypsum Gap	1955	
60	Pine Mountain	1955	
61	Calamity Mesa	1955	
64	Horse Range Mesa	1955	
65	Naturita N W	1955	
66	Joe Davis Hill	1955	
68	Egnar	1955	
69	Hamm Canyon	1955	
71	Davis Mesa	1955	
72	Paradox	1955	
77	Anderson Mesa	1955	(1956)
78	Uravan	1955	
81	Juanita Arch	1955	
83	Roc Creek	1956	
103	Golden	1957	
151	Louisville	1961	
152	Ouray	1962	
267	Central City	1964	
291	Ironton	1964	
397	Fort Lupton	1965	
398	Hudson	1965	
399	Platteville	1965	
486	Montrose	1965	(1966)
504	Telluride	1966	
512	Marble	1966	
536	Dolores Peak	1966	
578	Oh-be-joyful	1967	
596	Mount Tyndall	1968	
631	Bristol Head	1967	
702	Elk Springs	1968	
703	Banty Point	1968	
704	Chair Mountain	1968	
725	Hanover NW	1968	

Numerous localities for collecting fossils are mentioned in such publications as these of Junius Henderson:

"Paleontology of the Boulder Area," Colorado University Studies, vol. 2, 1904, p. 95-106.

"The Cretaceous Formations of Northeastern Colorado and

the Foothills Formation of North-Central Colorado," Colorado Survey Bulletin 19, 1920.

Enough information is given to enable one to find many of the original localities even today, such as "the brick yard, east of the University Hospital," "the field west of Haystack Butte," "on the Mesa south of the Chatauqua Grounds" (these are in the Boulder area), and others elsewhere.

There are a number of Fossil Creeks and similar geographic designations in Colorado that suggest likely places to look. For example, one of the most productive localities in Colorado has been Fossil Ridge (also known as Fossil Creek), 7 miles south of Fort Collins; this low, north-trending ridge of Hygiene sandstone — first extensively explored in 1906 by a University of Colorado expedition led by Professor Henderson — contains abundant, fossil-bearing concretions up to 4 feet in diameter, in which the bivalve *Inoceramus oblongus*, 6-8 inches long, is most common (Ref. 1).

Another example is Baculite Mesa, built of the Pierre formation and capped by the Nussbaum formation. It is situated east of U. S. 85-87, 7.2 miles north of the Pueblo city limit, and is named from its abundance of *Baculites*. These are extinct Cretaceous ammonoid cephalopods having straight (uncoiled) shells.

And of course there are many more such places in the state.

In 1930, J. Harlan Johnson (Ref. 2) summarized the fossil finds that had been made in the Denver area, coinciding with the Denver quadrangle of the U. S. Geological Survey.

A catalog of the type specimens of fossils in the University of Colorado Museum was published in 1938 by Hugo G. Rodeck (Ref. 3). Accounts of acquisitions made during field work and by other means have appeared in the University of Colorado Studies and in the Annual Reports of the Colorado (since 1948, the Denver) Museum of Natural History.

George P. Merrill (Ref. 4) prepared a "Catalogue of the Type and Figured Specimens of Fossils, Minerals, Rocks and Ores in the Department of Geology, United States National Museum." The American Museum of Natural History, in New York, and other museums have issued similar catalogs in which Colorado specimens are well represented.

Numerous, miscellaneous listings of fossil localities in Colorado, taken mostly from the older literature, are given in *Fossils in America,* by Jay Ellis Ransom (Harper & Row, Publishers, New York, 1964).

A somewhat similar book is *Hunting for Fossils,* by Marian Murphy (The Macmillan Company, New York, 1967); it is less complete but easier reading.

The only book besides the present one that emphasizes Colorado occurrences is *Fossils; A Story of the Rocks and Their Record of Prehistoric Life,* by Harvey C. Markman (Denver Museum of Natural History, 4th edition, 1961).

The general, popular literature on fossils is fairly extensive. Useful, classified and annotated lists are given in these two publications:

Fossils. An Introduction to Prehistoric Life, by William H. Matthews III, Barnes and Noble, Inc., New York, 1962.

Earth for the Layman, by Mark W. Pangborn, Jr., American Geological Institute, Washington, 2d edition, 1957.

The best of the most accessible, inexpensive books on fossils at the beginning level are the following:

Fossils; An Introduction to Prehistoric Life, by William H. Matthews III, Barnes and Noble, Inc., New York, 1962.

Fossils. A Guide to Prehistoric Life, by Frank H. T. Rhodes, H. S. Zim, and Paul Shaffer, Golden Press, New York, 1962.

Care should be taken not to destroy specimens for any reason. A large proportion of worthwhile finds of fossils have been made by amateurs, who have likewise accounted for a large proportion of unnecessary damage to valuable material. You can be an amateur scientist or a professional vandal but not both. Private and public rights have priority over your own. Instructions for recovering and preparing specimens are given in the book *Fossils; An Introduction to Prehistoric Life,* by William H. Matthews III (Barnes and Noble, Inc., New York, 1962). His advice on where to collect and where not to is worth reading.

REFERENCES

Ref. 1. University of Colorado Studies, vol. 5, 1908, p. 179-192.
Ref. 2. Colorado Scientific Society Proceedings, vol. 12, 1930, p. 355-376.
Ref. 3. University of Colorado Studies, vol. 25, 1938, p. 281-304.
Ref. 4. U. S. National Museum Bulletin 53, pt. 1, 1905; pt. 2, 1907.

Chapter 22

Fossil Plants

Vegetation being necessary for the existence of animal life, the plant fossils of Colorado are discussed here first, just as they appeared first in the geologic history of the state. In the form of petrified wood — including gemmy agate and valuable, uranium-filled logs — they include some of the most interesting of Colorado fossils.

ALGAL LIMESTONE

The oldest plants are the thallophytes, which comprise the algae (including diatoms) and the fungi, and both algae and fungi living together as lichens. These plants, the simplest of which are one celled and microscopic, lack roots, stems and leaves; they are abundant in certain deposits in Colorado. They are most conspicuous in the so-called algal limestones, which may develop as large reefs and in places may contain petroleum or ore minerals. Oil in the Green River formation of northwestern Colorado is an example. Rounded, banded masses of stromatolite algae are known as algal biscuits.

J. Harlan Johnson, of the Colorado School of Mines, has specialized in the subject of algal limestones and has written extensively on them. His recent book *Limestone-Building Algae and Algal Limestones* (Colorado School of Mines, Golden, 1961), contains some photographs of Colorado specimens. Many of the so-called coral reefs of Colorado's past history, like those of today's tropic seas, have been built chiefly by organisms other than the corals themselves; algae are often the predominant contributors. Dr. Johnson has described important algal limestones in South Park of Permian age (Ref. 1), Pennsylvanian age (Ref. 2), and Oligocene age (Ref. 3); on the west side of the Sawatch Range, as far as Glenwood Springs, of Mississippian age (Ref. 4); and elsewhere in central Colorado.

An extensive flora of very minute algae and related (cryptogamic, or spore-reproducing) plants occurs in the oil shales of the Green River formation of northwestern Colorado, as first described in 1916 by Charles A. Davis (Ref. 5) and more fully treated by F. H. Knowlton (Ref. 6) and W. H. Bradley (Ref. 7-8); related references are given by Roland W. Brown (Ref. 9). Microscopic plants and their parts (seeds, spores, pollen) are valuable to the oil geologist and micropaleontologist for the information they give about geologic history, but they are of little interest to the usual collector of fossils. Fine, large plant fossils also occur in the Green River beds, as illustrated by Dr. Brown.

Reefs of algal limestone as thick as 18 feet are abundant in the Green River formation. Dr. Bradley (Ref. 7) describes their significance as indicating the shallow, clear nature of the parts of the lake in which they were formed, the presence of vegetation in sheltered bays, and the availability of dissolved calcium salts in the water.

HIGHER PLANTS

Although the first American fossil cycad to be discovered was found about 1860 between Baltimore and Washington, the first to be described came from Colorado in 1876. Cycads are tropical, palmlike plants reproducing by means of spermatozoids.

In spite of the relative scarcity of Carboniferous plants in the western part of the United States, in comparison with the profusion known in the coal beds of the eastern and central states, more have been described from Colorado than from any other state west of the Great Plains. Leo Lesquereux (Ref. 10) described the first ones, from near Fairplay, in 1882.

The discovery in 1953 by G. Edward Lewis of "palmlike leaves" from a primitive fossil palm in the Dolores formation of

157

Triassic age, near Placerville, in San Miguel County, was highly significant. The name *Sanmiguelia lewesi* was given it by Roland W. Brown (Ref. 11) and represents the oldest fossil angiosperm, or flowering plant, known anywhere in the world.

The great Coal Measures of the Paleozoic Era in the eastern and central parts of the United States had their counterpart in Colorado during the last part of the Mesozoic Era. The fossil plants of the coal beds and associated formations have been described at length in major publications of the U. S. Geological Survey and its predecessors, chiefly by Leo Lesquereux, F. V. Hayden (Ref. 12), John Strong Newberry (Ref. 13), and F. H. Knowlton (Ref. 14-16).

Plant fossils in the Laramie and Denver formations outcropping at the surface near Denver represent many ferns and shrubs, and such trees as palm, breadfruit, poplar, oak, laurel, and especially fig. The fossil plants of the Denver Basin were summarized by F. H. Knowlton in the classic monograph on that area by Samuel Franklin Emmons, Whitman Cross, and George Homes Eldridge (Ref. 17). John L. Le Conte is credited with making the first collection of them in 1867 while attached as geologist to the survey for the extension of the Union Pacific Railway from Kansas to the Rio Grande. J. Harlan Johnson (Ref. 18) described abundant plant fossils of about 225 species along the south side of South Table Mountain and the northwestern foot of Green Mountain, near Golden.

AMBER

The Laramie formation at Marshall, in Boulder County, has been an important coal producer for the city of Boulder and yielded important plant fossils to Leo Lesquereaux. In 1909, T. D. A. Cockerell (Ref. 19) found translucent, orange-brown "amber" in the coal. Described as becoming opaque and friable in ether, it is thus not true amber, which contains succinic acid and does not react to ether. Similar, resinous material has since been found elsewhere in Colorado. Amber with an imprisoned insect was found near Golden, as reported by J. Harlan Johnson (Ref. 20). Amber has been found with jet in the Pikes Peak region by Robert D. Wilfley.

Petrified wood — also referred to as silicified, agatized, jasperized, or opalized wood — is widespread in Colorado, occurring in rocks of various ages, dating from the first appearance of land plants during the Silurian Period. A number of good collecting localities for gem material, notably in the east-central counties of the state, are described in *Colorado Gem Trails and Mineral Guide*. Wood has also been replaced by many other minerals besides quartz and opal.

URANIUM IN PETRIFIED WOOD

The association of uranium mineralization with petrified wood is familiar in parts of Colorado where deposits have been mined for their uranium content.

Most of the uranium-soaked fossil wood occurs in ancient stream channels, especially at bends and junctions. In size, it varies from tiny shreds to complete logs and includes ferns and leaves. The torn and shredded plant debris is referred to as "carbon trash." Uranium is not so likely to be found in silicified wood as in the carbonized material. The distinctive "roll deposits" of uranium mineralization in the Colorado Plateau — most common in the Salt Wash sandstone — sometimes have a core of petrified wood.

A famous example of the rich mineralization occasionally found in Colorado Plateau fossil wood was reported in 1933 by Frank L. Hess at the Cracker Jack mining claim, Calamity Gulch. A total value of $350,000 (in 1920 prices) was obtained from the uranium, vanadium, and radium in two logs and the sandstone lying between them. Two other logs, on the Dolores property, brought $232,900. A content of 20 percent uranium oxide (U_3O_8) was found in similar fossil logs at the Maggie C deposit, in Long Park.

PETRIFIED FORESTS

Perhaps no other natural exhibit of fossils in the state of Colorado can equal the standing petrified tree stumps near Florissant, in Park County. This is a scenic and scientific treat of exceptional merit, well worth the admittance fee to those interested in ancient life. Nearest Florissant, 1.8 miles south on Colorado 143, is the Colorado Petrified Forest; at 2.4 miles is the Pike Petrified Forest.

The Trio, in the second forest, is a cluster of three *Sequoia* stumps of similar size and appearance, reaching a height of 13.5 feet. Each stump is 6 feet in diameter and is bound by iron cables; the whole group would require a rope 85 feet long to encircle it. The roots are said to be entangled in a single, perhaps unique root system.

Standing elsewhere is the largest petrified tree ever recorded: 55 feet in circumference, 140 tons in weight, and 10.5 feet in height. Like the rest, it is disintegrating in the weather. One tree that has already fallen apart has given rise in its midst to a living pine.

At least three distinct layers of the petrified trees can be seen beneath the layer of the Florissant lake beds that contain the most leaves, seeds, and insects. First is a thin bed that contains scattered petrified trunks. Below it is a zone containing trunks up to 5 feet in diameter; most of these are lying down and have been flattened. Underneath is a brown volcanic shale, weathered gray, toward the base of which are the large stumps and trunks that are found standing erect and in place. The species of *Sequoia* is more closely related to the California coast redwood of today than to the Sierra redwood.

FLORISSANT FOSSIL BEDS

One of the most outstanding fossil localities in the world surrounds the village of Florissant, in Park County, close to the geographic center of Colorado. From volcanic beds of so-called "paper shale" deposited here in shallow and temporary lakes during the Oligocene Epoch have been taken a wealth of plants, insects, fish, and other fossils that are to be seen in museums and private collections throughout America and Europe. More than half of all known fossil butterflies have come from this one small area, as well as other interesting insects. Mollusks, birds, and

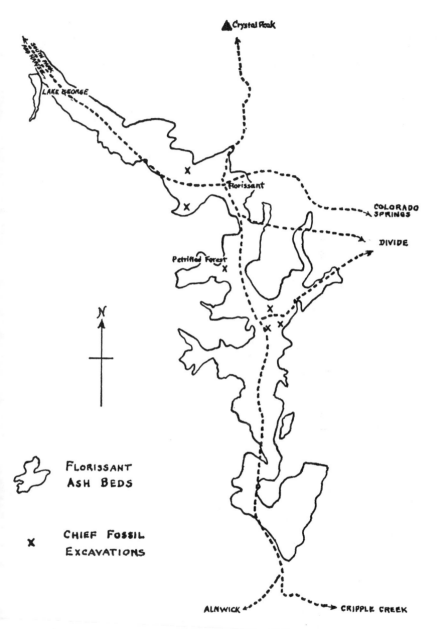

Crystal Peak

SOUTH PARK
AND DENVER

LAKE GEORGE

×

Florissant

COLORADO SPRINGS

DIVIDE

Petrified Forest
×

N

×
×

FLORISSANT
ASH BEDS

X CHIEF FOSSIL
EXCAVATIONS

ALNWICK

CRIPPLE CREEK

Map of Florissant Fossil Beds. *Harry D. MacGinitie, Carnegie Institution of Washington.*

161

mammals are also among the fossils here. The number of species of fossil plants is about 125. Nearly every textbook on historical geology describes this remarkable, Pompeii-like preservation, buried in rains of ash during repeated eruptions of an ancient volcano. This may be the most highly fossiliferous deposit on the planet Earth.

As part of the work of the historic Hayden Geological Survey — the United States Geological and Geographical Survey of the Territories — A. C. Peale first discovered the Florissant beds in 1873. Samuel H. Scudder collected many specimens here and wrote extensively about them, as have others since then, especially T. D. A. (Theodore Dru Alison) Cockerell, of the University of Colorado, who virtually adopted the locality as his own and published about it at frequent intervals for 30 years.

Seven sedimentary and volcanic layers, each of varying thicknesses, have been described here by Harry D. MacGinitie (Ref. 21). Of these, only the one containing plant fossils was deposited in ancient Lake Florissant, as the irregularly shaped body of water has been named. Falling ash brought down insects and birds, trapping them with fish, other animals, and plants, all in an extraordinary assemblage. The rest of the beds are mudflows and reworked volcanic ash that was laid down in a river. The source of the volcanic activity seems to have been about 15 miles away, near Guffey. The present exposures of the strata, as shown on the accompanying map, are the result of subsequent faulting, folding, and erosion.

The best preserved and most numerous impressions of leaves, seeds, and insects occur in the paper shale of the bottom half of the lake beds. The crumbly shale can be preserved by coating it with shellac or a similar substance. Beneath this paper shale, fossil wood is common in the southern part of the present basin; most of it consists of large stumps and trunks of *Sequoia* (see page 160).

Associated with *Sequoia* are fossils of pine, spruce, fir, cottonwood, juniper, alder, oak, rose, ash, grape, maple, sumac, thorn, thistle, aster, and Virginia-creeper — all still growing in Colorado. Also found, but no longer living in Colorado, are chestnut, fig, basswood, soapberry, sweet firn, sweet gum, holly, smoke tree, and persimmon.

The combined evidence of flora and fauna point to a river-bottom environment interspersed with shallow lakes or ponds in a warm-temperate climate having a pronounced dry season. Thus saith MacGinitie.

Negotiations are currently under way for the establishment of a national monument in the Florissant area, to embrace both the so-called lake beds and the petrified forests.

Fossil Plants of the Florissant Beds, Colorado, by Harry D. MacGinitie (Carnegie Institution of Washington Publication 599, Washington, 1953) — one of a series of important Contributions to Paleontology — is the most recent comprehensive book on the Florissant locality. Although it deals principally with the flora, in 198 pages and 75 plates, the geography and geology are summarized. A bibliography of 209 entries is given, many of which are concerned with other places than Florissant.

The latest geologic study of Florissant is by Carol Ann McLeroy and Roger Y. Anderson *(Geological Society of America Bulletin,* volume 77, 1966, pages 605-618).

A very interesting, popular article on Florissant is "An Insect Pompeii," by Reginald D. Manwell (*Scientific Monthly,* volume 80, 1955, pages 356-361).

CREEDE FOSSIL BEDS

Plant and insect fossils different and younger in age than those at Florissant but occurring in similar, readily separated, thin layers of volcanic shale are found near Creede, in Mineral County. The beds here belong to the Pliocene Epoch and are said to contain the only montane fossil flora in North America, having much in common with the living plants of the Front Range at elevations between 6,000 and 8,000 feet (Ref. 21, p. 73).

The Creede deposits, in the Creede formation, were first described by F. H. Knowlton (Ref. 22) in 1923, although collecting had been done earlier by others. These rocks, which are mostly rhyolite tuff, were deposited in a deep lake basin having steep sides. The fossil plants contain many representatives of conifers and lesser amounts of poplar, alder, and shrubs and vines, which include currant, grape, and especially Oregon grape. Most of the fossils come from outcrops along the Rio Grande and its tributaries in the vicinity of Creede. A good locality, according to Esper S. Larsen, Jr. and Whitman Cross (Ref. 23), is in cliffs on the north bank of the Rio Grande, above Sevenmile Bridge, which is at Marshall Park campground, southwest of Creede.

F. H. Knowlton prepared "A Catalogue of the Mesozoic and Cenozoic Plants of North America" (Ref. 24). The localities are given for many Colorado plants, which are also listed by formation and age, grouped according to general (but not detailed) localities.

In addition to the textbooks on historical geology, information on plant fossils is given in the following books:

Ancient Plants and the World They Live In, by Henry N. Andrews, Jr., Comstock, Ithaca, New York, 1947. A popular survey.

An Introduction to Paleobotany, by Chester A. Arnold, McGraw-Hill Book Company, New York, 1947. A college textbook, illustrated.

Plants of the Past; A Popular Account of Fossil Plants, by Frank Hall Knowlton, Princeton University Press, Princeton, N. J., 1927.

Principles of Paleobotany, by William C. Darrah, The Ronald Press Company, New York, 2d edition, 1960. A college textbook.

Studies in Paleobotany, by Henry N. Andrews, Jr., John Wiley and Sons, Inc., New York, 1961. A college textbook.

Tree Ancestors; A Glimpse into the Past, by Edward Wilber Berry, Williams and Wilkins Company, Baltimore, 1923. An interesting book.

REFERENCES

Ref. 1. Bulletin of the American Association of Petroleum Geologists, vol. 17, 1933, p. 863-865.

Ref. 2. Geological Society of America Bulletin, vol. 51, 1940, p. 571-595.

Ref. 3. Geological Society of America Bulletin, vol. 48, 1937, p. 1227-1235.

Ref. 4. Geological Society of America Bulletin, vol 56, 1945, p. 829-847.

Ref. 5.   Proceedings of the National Academy of Sciences, vol. 2, 1916, p. 114-119.
Ref. 6.   U. S. Geological Survey Professional Paper 131-F, 1923, p. 133-182.
Ref. 7.   U. S. Geological Survey Professional Paper 154-G, 1929, p. 203-223.
Ref. 8.   U. S. Geological Survey Professional Paper 168, 1931.
Ref. 9.   U. S. Geological Survey Professional Paper 185-C, 1934.
Ref. 10.   Harvard College Museum of Comparative Zoology Bulletin, vol. 7, Geological Series, no. 1, 1882, p. 243-247.
Ref. 11.   U. S. Geological Survey Professional Paper 274-H, 1956, p. 205-209.
Ref. 12.   U. S. Geological and Geographical Survey of the Territories Report, vol. 7, 1878.
Ref. 13.   U. S. Geological Survey Monograph 35, 1898.
Ref. 14.   U. S. Geological Survey Professional Paper 101, 1917, p. 223-435.
Ref. 15.   U. S. Geological Survey Professional Paper 130, 1922.
Ref. 16.   U. S. Geological Survey Professional Paper 155, 1930.
Ref. 17.   U. S. Geological Survey Monograph 27, 1896, p. 466-473.
Ref. 18.   Colorado Scientific Society Proceedings, vol. 12, 1930, p. 355-376.
Ref. 19.   Torreya, vol. 9, 1909, p. 140-142.
Ref. 20.   Colorado Scientific Society Proceedings, vol. 12, 1930, p. 355-376.
Ref. 21.   Carnegie Institution of Washington Publication 599, 1953.
Ref. 22.   U. S. Geological Survey Professional Paper 131-G, 1923.
Ref. 23.   U. S. Geological Survey Professional Paper 258, 1956.
Ref. 24.   U. S. Geological Survey Bulletin 696, 1919.

Chapter 23

Invertebrate Fossils

Ranging in size from microscopic, one-celled protozoans to the giant squid, the animals grouped together as invertebrates are those that lack a backbone. They represent more than 97 percent of all animal species. (Some biologists put the protozoans, bacteria, and certain other low organisms, together with the one-celled plants previously described, into the Kingdom Protista — neither animal nor plant.) The invertebrates are divided into about 30 phyla, of which 12 are common fossils. Among these phyla are the porifera (sponges), coelenterates (corals, jellyfish, etc.), bryozoans (moss animals), brachiopods (bivalve, shelled animals), mollusks (shellfish), arthropods (jointed animals), and echinoderms (starfish, etc.).

The most conspicuous Colorado fossil invertebrates, except perhaps the fossil insects at Florissant, date from the Mesozoic Era. The so-called Age of Invertebrates, by which paleontologists mean the Cambrian and Ordovician Periods of the Paleozoic Era, left only relatively meager fossils of such prominent forms as trilobites, brachiopods, mollusks, corals, and bryozoans — at least, in comparison with the rich deposits found elsewhere. The reason is that our Paleozoic rocks are thinner, more deformed, and less representative of the entire geologic column than in such states as New York, Ohio, and Illinois, for example.

An excellent collecting locality for Paleozoic invertebrate fossils — trilobites, brachiopods, and gastropods of Ordovician age — has been reported by N. W. Bass (Ref. 1) in Garfield County, on U. S. 6-24, between Walcott and Glenwood Springs. At 2.3

miles west of the county line and 0.2 miles west of the White River National Forest boundary sign is the outcrop: the rock slopes beneath a brown cliff of the Tie Gulch member of the Manitou formation.

Later Paleozoic rocks in Colorado are likewise deficient in major accumulations of invertebrate fossils, although some few exceptions are known. The Elbert formation, for example, has in places an abundance of invertebrate fossils of Devonian age, as well as interesting fish remains. Those of the Ouray formation of the same age, the first Devonian fossils found in Colorado (by F. M. Endlich in 1875), were described by E. M. Kindle (Ref. 2). The Mississippian and Pennsylvanian faunas were described in much detail by George H. Girty (Ref. 3), who gave a long "register of localities."

Again, after the Mesozoic Era had come and gone, no significant invertebrate fossils, apart from those at Florissant, were presumably left within the state during the Cenozoic Era. This is because the seaways, in which invertebrates thrive in such abundance, had retreated to the oceans and continental shelves.

MESOZOIC INVERTEBRATES

In Colorado, as well as the rest of the world, the dominant Mesozoic invertebrates were mollusks, or shellfish. These included cephalopods, pelecypods, and gastropods. After becoming almost extinct, corals expanded in the Mesozoic seas, as shown by coral reefs in the Morrison formation in Colorado.

The most important and interesting of the Mesozoic mollusks were the ammonites, whose chambered, coiled, marine shells are extremely common in Colorado rocks. These cephalopods, similar to the present-day pearly nautilus, had shells as large as 9 feet in diameter, and specimens 3-4 feet across are not uncommon, but 3-4 inches was about average size. Inside was a pearly lining. Outside, many were ornamented in a wide variety of ribs, spines, and knobs. The particular pattern of the sutures — which mark the lines of attachment of the thin, pearly partitions (septa) to the shell itself — identifies the kind of ammonite and hence the age of the sedimentary rock it is found in. Some sutures are

167

very complex, and some ammonite shells are very beautiful. Ammonites became entirely extinct at the end of the Mesozoic Era.

Another important Mesozoic cephalopod in Colorado is named belemnite. This animal had an internal shell shaped like a cigar; its appearance was like that of a modern squid. Like the squid, also, it had an ink sac, which served for protection. The Sundance formation, in Routt County, is particularly rich in belemnite fossils.

Popularly thought of as clams, these bivalve mollusks called pelecypods are abundant as fossils in Colorado rocks of marine origin. The two halves of the shell are similar and were attached to the right and left sides of the animal, opening along a rather straight hinge line, the structure of which is used to decide the kind and age of the specimen. The shell is composed of three layers, the inner one being mother-of-pearl.

Familiarly known as snails, gastropods are mostly aquatic mollusks. Their coiled shells are numerous in Colorado rocks. From the Morrison formation, of Jurassic age, Teng-Chien Yen (Ref. 4) described and illustrated 32 fresh-water species (19 genera) from 42 historic localities in Colorado and nearby states, as well as 17 species of pelecypods (3 genera) and some ostracodes and sponge spicules. More nonmarine mollusks of the same classes but of Cretaceous age were described from the Book Cliffs coalfield area of western Colorado by the same author (Ref. 5).

An interesting fossil is the occurrence at Aspen of gastropods completely replaced by native silver, and others by sphalerite and galena.

TEPEE BUTTES

Conical hills, ranging in height up to about 50 feet, are a conspicuous feature of the landscape in south-central and east-central Colorado, where they stand in scattered rows above the surface of Pierre shale. Because of their resemblance to Indian tents, they are known as Tepee Buttes. They are almost solid masses of marine fossils, especially the pelecypod *Lucina occidentalis*. In the vast Cretaceous seas that formerly covered this

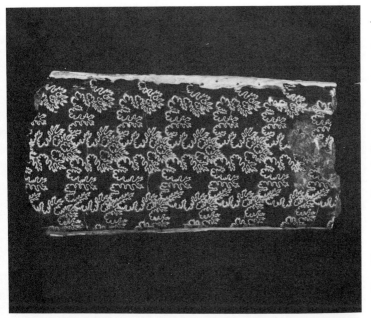

Baculite ammonite of Cretaceous age, showing sutures. *Ward's Natural Science Establishment.*

Crinoid of Mississippian age, an animal that looks like a plant. *Ward's Natural Science Establishment.*

World's largest petrified tree stump, Florissant, Colorado. *George L. Beam.*

Tribolite of Cambrian age. *Ward's Natural Science Establishment.*

Brachiopod of Devonian age. *Ward's Natural Science Establishment.*

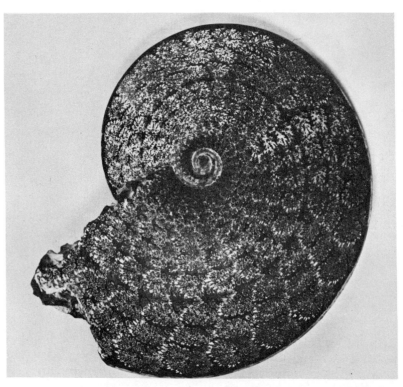

Ammonite of Cretaceous age. *Ward's Natural Science Establishment*

Algal reef wth layers of pyrite. *W. H. Bradley, U. S. Geological Survey.*

The newest mounted dinosaur from Colorado. This is *Haplocanthosaurus* from near Canon City. *Cleveland Museum of Natural History.*

Quarry Visitor Center at Dinosaur National Monument, Colorado. *National Park Service.*

Revealing dinosaur vertebra at Dinosaur National Monument, Colorado. *Dinosaur Nature Association.*

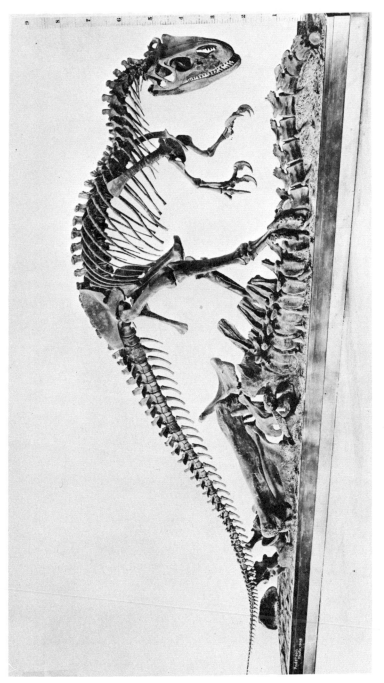

Professor Cope's *Antrodemus* dinosaur from Garden Park, Colorado. *American Museum of Natural History.*

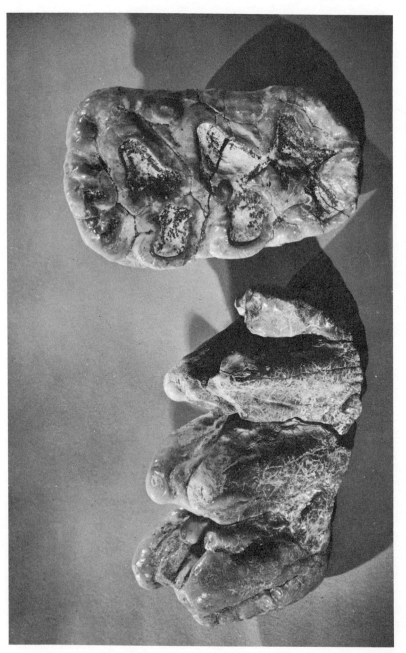

Mastodon molar teeth of Pleistocene age. *Ward's Natural Science Establishment.*

region, these mounds were apparently reef-like structures growing where the temperature, clarity of water, and food supply were most favorable.

On the geologic map of Colorado, the Tepee Buttes are shown as black dots. Some of the Tepee Buttes between Colorado Springs and Pueblo, along U. S. 85-87, can be easily located according to the following road log by John Chronic (Ref. 6), beginning at 0.0 where the exit road goes off to the town of Fountain:

1.6	To the east and south.
2.3	To the south, on the east side of Fountain Creek.
6.1	To the east.
11.3	To the east, on the horizon.
12.1	On the east side of Fountain Creek.
14.2	To the southeast.
16.6	To the east, on the horizon.

The Tepee Butte nearest Colorado Springs is situated east of Colorado 115, 1.5 miles south of the "braided intersection," where Colorado 115 begins.

INSECTS

The most remarkable of the insect fossils of Colorado are undoubtedly the butterflies of Florissant. These were first described in detail by Samuel H. Scudder (Ref. 7) and have attracted the attention of entomologists ever since. Richard M. Fox, formerly at Colorado College, believes that the mystery of the present-day distribution of tropical butterflies in Central America might be solved by further study of these ancient insects. They were perhaps first killed by noxious gases and then downed in a rain of volcanic ash above Lake Florissant between 25 and 40 million years ago.

Henry Fairfield Osborn regarded the tsetse fly at Florissant, and the horse fly also found fossilized here, as suggesting the possible extinction of some of the American mammals by epidemic, a view not much in vogue now. Other insects — mostly extinct kinds — include grasshoppers, dragon flies, locusts, and plant lice. Ants are especially common, representing about one-

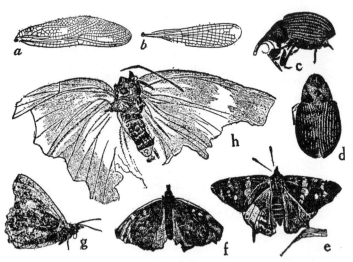

Florissant Insect Fossils: *a, b Dragonflies; c, d Beetles; e-h Butterflies, moths.*

quarter of the total; wasps and bees constitute another 15 percent. True flies total 30 percent; beetles and weevils, 13 percent; and true bugs, 11 percent, according to Dr. Scudder.

The more important references on Florissant insects include especially those by Samuel Hubbard Scudder.

Dr. Scudder (Ref. 8) recovered and described cockroaches from Triassic beds near Fairplay, made known to him by Arthur Lakes, professor at the Colorado School of Mines. Dr. Scudder (Ref. 9) also described fossil ants from South Park. A new locality for Oligocene fossil insects was discovered in 1962 by Christopher J. Durden, in Chase Basin, Park County (Ref. 10).

The enormous literature on invertebrate fossils is being compactly summarized in the 24 volumes of the *Treatise on Invertebrate Paleontology,* edited by Raymond C. Moore (Geological Society of America and University of Kansas Press, 1953-).

Besides from textbooks on historical geology, knowledge of invertebrate fossils can conveniently be obtained from the following books:

Animals Without Backbones; An Introduction to the Invertebrates, by Ralph Buchsbaum, University of Chicago Press, 2d edition, 1948. An interesting textbook.

Index Fossils of North America, by Hervey W. Shimer and Robert R. Shrock, John Wiley and Sons, Inc., New York, 1944. The major reference key to invertebrate fossils of the continent, profusely illustrated.

Invertebrate Paleontology, by W. H. Easton, Harper and Row, New York, 1960. A college textbook, well illustrated.

Invertebrate Fossils, by Raymond C. Moore, Cecil G. Lalicker,, and Alfred G. Fisher, McGraw-Hill Book Company, New York, 1952. A well-illustrated college textbook.

Principles of Invertebrate Paleontology, by Robert R. Shrock and William H. Twenhofel, McGraw-Hill Book Company, New York, 1953. A college textbook, well-illustrated.

Search for the Past; An Introduction to Paleontology, by James R. Beerbower, Prentice-Hall, Inc., Englewood Cliffs, N. J., 2d edition, 1968. Covers both vertebrate and invertebrate fossils.

REFERENCES

Ref. 1.   Guide to the Geology of Colorado, Geological Society of America *et al,* Denver, 1960, p. 79.

Ref. 2.   U. S. Geological Survey Bulletin 391, 1909.

Ref. 3.   U. S. Geological Survey Professional Paper 16, 1903.

Ref. 4.   U. S. Geological Survey Professional Paper 223-B, 1952, p. 21-51.

Ref. 5.   U. S. Geological Survey Professional Paper 254-B, 1954, p. 59-66 and plates.

Ref. 6.   Guide to the Geology of Colorado, Geological Society of America *et al,* Denver, 1960, p. 100-117.

Ref. 7.   U. S. Geological Survey Eighth Annual Report, 1886-87 (1889), p. 433-474 and 2 plates.

Ref. 8.   Academy of Natural Sciences of Philadelphia Proceedings, 1885 (1886), p. 34-39, 105-115.

Ref. 9.   American Naturalist, vol. 11, 1877, p. 191.

Ref. 10. Journal of Paleontology, vol. 40, 1962, p. 215-219.

Chapter 24

Vertebrate Fossils

These are the most spectacular of fossils, in Colorado as elsewhere, for they range in size up to the great dinosaurs, and in complexity up to man. Fossil man — a newcomer to Colorado — is, however, not discussed here; for this story, you are referred to the authoritative book *Ancient Man in North America,* by H. M. Wormington (Denver Museum of Natural History, 4th edition, 1957). Among the Tertiary primates of Colorado was the oldest fossil monkey in America, described from Miocene rocks of unidentified locality in 1873 by E. D. Cope (Ref. 1) at the Academy of Natural Sciences of Philadelphia.

FOSSIL FISH

The backboned animals made their appearance with the fish, which represent four classes of vertebrates. Few of today's fishermen who consider Colorado an excellent state in which to enjoy their sport realize that, as far as is now known, the first fish in the world lived in Colorado. At least, the oldest unquestioned remains of fish — and of any vertebrate, except a single small jaw found in Missouri — have been taken from the Harding sandstone of Ordovician age, 1 mile northwest of the State Penitentiary, at Canon City. Although they are only small bony plates and scales, the evidence is conclusive: fish were residents in Colorado about as early as can yet be proved for any other place on earth. Although fish of the same age have since been found elsewhere in Colorado and in other places, the next younger ones came 50 million years later than ours. Two paleontologists named Timothy W. Stanton and Israel C. Russell found the first specimens, and Charles D. Walcott (Ref. 2) made the identification.

These jawless fish, *Astraspis desiderata* and *Eriptychius americanus*, of the class Agnatha, are called ostracoderms. They seem to have lived in fresh-water streams and lakes, or in estuaries. The forward part of the body was armored against the appetite of the eurypterids, which apparently fed on them, for when the ostrocoderms declined, so did the eurypterids. (Ref. 3-4).

Since the Devonian Period, in the second half of the Age of Fishes, sharks have been dread denizens of the seas. Bcause this fierce animal has a cartilaginous skeleton, the only parts suitable for preservation as fossils are the teeth and some spines. Shark's teeth are familiar to fossil collectors in certain places in Colorado, such as in hogbacks of the Cretaceous formations along the Front Range.

REFERENCES

Ref. 1. Academy of Natural Sciences of Philadelphia Proceedings, 1873, p. 419.
Ref. 2. Geological Society of America Bulletin, vol. 3, 1892, p. 153-172.
Ref. 3. Geological Society of America Proceedings, 1937 (1938), p. 289-290.
Ref. 4. Bulletin of the American Association of Petroleum Geologists, vol. 38, p. 284-305.

Chapter 25

Fossil Reptiles

With the evolution of the reptiles from the amphibians — which are unimportant as Colorado fossils — completely land-dwelling vertebrates appeared. Although released from the amphibians' dependence upon the water as the place to lay eggs, some of the reptiles returned secondarily to the water as swimming forms in the Mesozoic Era, the Age of Reptiles. Throughout the 130 million years of this medieval era of life on earth, dinosaurs and the fellow reptiles reigned supreme on land, in the sea, and in the air. These animals were cold-blooded, mostly egg-laying, vertebrates having dry, scaly skins; thus, they were like the living reptiles of today — the snakes, lizards, turtles, and crocodiles, and the curious tuatara of New Zealand. The modern crocodile is more closely related to the dinosaurs than is any other animal.

DINOSAURS

The dinosaurs comprise two distinct biologic orders of reptiles, differing mainly in the structure of the hip bones. Although the word Dinosauria, coined in 1842 by Sir Richard Owen, means "terrible lizards," some dinosaurs were as small as a chicken. Most, however, were of gigantic size. Some were carnivorous, others herbiverous. Some walked on two legs; others, developing later, walked on four. Many stayed on land, and many others dwelled in swamps and marshes. Below is a diagram showing the classification of the dinosaurs.

Order Saurischia
(saurichians)
Suborder Theropoda
(theropods)
Suborder Sauropoda
(sauropods)

Order Ornithischia
(ornithischians)
Suborder Ornithopoda
(ornithopods)
Suborder Stegosauria
(stegosaurs)
Suborder Ankylosauria
(ankylosaurs)
Suborder Ceratopsia
(ceratopsians)

The saurischian dinosaurs, which had hip bones typically reptilian, were the first to evolve from the thecodonts — the ancestors of the dinosaurs, flying reptiles, crocodiles, and eventually the birds. They included *Brontosaurus, Diplodocus,* and *Brachiosaurus,* all of which were huge plant-eaters living in Colorado. The saurischians also included the flesh-eating *Ceratosaurus* and culminated in *Tyrannosaurus,* the most ferocious carnivore of all times.

The ornithischian dinosaurs, which had a hip structure like that of birds, came on the scene later than the first saurischians and were more advanced in their evolution. They included four principal types: the ornithopods (various two-footed dinosaurs, such as the duck-billed dinosaur *Trachodon,* with its webbed feet and its 2,000 grinding teeth); the plated dinosaurs (such as *Stegosaurus*); the armored dinosaurs (such as *Ankylosaurus*); and the horned dinosaurs (such as *Protoceratops* and *Triceratops,* the former of which was responsible for the dinosaur eggs first discovered in Mongolia by the Roy Chapman Andrews expedition in 1922). All these dinosaurs were vegetarians. *Stegosaurus* is especially noted for its "two sets of brains" — actually enlargements of the spinal-nerve cord in the hips and shoulders;

these swellings were much larger than the brain itself and controlled the movements of the legs and tail. This odd development inspired Bert Leston Taylor of the *Chicago Tribune* to write the following verse:

> Behold the mighty dinosaur,
>> Famous in prehistoric lore,
> Not only for his power and strength
>> But for his intellectual length.
> You will observe by these remains
>> The creature had two sets of brains—
> One in his head (the usual place),
>> The other at his spinal base.
> Thus he could reason "A priori"
>> As well as "A posteriori."
> No problem bothered him a bit,
>> He made both head and tail of it.
> So wise was he, so wise and solemn,
>> Each thought filled just a spinal column.
> If one brain found the pressure strong
>> It passed a few ideas along.
> If something slipped his forward mind
>> 'Twas rescued by the one behind.
> And if in error he was caught
>> He had a saving afterthought.
> As he thought twice before he spoke
>> He had no judgment to revoke.
> Thus he could think without congestion
>> Upon both sides of every question.
> Oh, gaze upon this model beast,
>> Defunct ten million years at least.

Stegosaurus is likewise famous for the double row of big, vertical plates running along its back, for its remarkably small head, and for the two pairs of sharp spikes at the end of its tail.

The first Colorado dinosaur was discovered in 1876. What might have been a dinosaur bone was reported in 1787 in New Jersey. Dinosaur footprints, for a long while thought to be bird tracks, were found in Massachusetts in 1802; and "human remains" recovered there in 1818 proved later to have been dinosaur fossils. Captain William Clark, of the Lewis and Clark expedition, probably had a dinosaur bone when he mentioned in his journal "the rib of a fish" he had pried loose in Montana in 1806. Dr. Ferdinand V. Hayden collected dinosaur teeth and bones in Montana and South Dakota in 1855.

Agatized dinosaur bone, often very attractive as a novel gem stone, comes from many places in western Colorado, from Garden Park, and in eastern Colorado, especially south of La Junta. This material is being carved in West Germany as miniature turtles, because its cell structure resembles the pattern of the animal's shell.

MORRISON DINOSAURS

The year 1877 was marked in Colorado by the discovery of the burial places of dinosaurs in two of the main localities of the world where these prehistoric animals have become known. One of the localities is west of Denver, between Golden and Morrison; the other is Garden Park, north of Canon City. These, together with Como Bluff, Wyoming — also discovered in 1877 — represent a related series of dinosaur deposits in the Morrison formation, which yielded a large amount of specimen material and priceless information about the great reptiles whose remains it was. The Morrison also yields pterodactyls, turtles, crocodiles, fishes, birds, primitive mammals, invertebrates, and plant fossils.

In March 1877, Arthur Lakes, a former English clergyman who was teaching at the "State School of Mines," and H. C. Beckwith found a dinosaur vertebra and a femur 14 inches across in the hogback near Morrison. When three-quarters of a ton of petrified bones reached Othniel Charles Marsh at Yale College, he employed Professor Lakes and Benjamin F. Mudge as fossil collectors, and they gathered another ton and a quarter of bones in about 2 weeks. Joined awhile by Captain Beckwith in 1877, Professor Lakes continued to collect until the summer of 1879.

On this same hogback, he "trotted out the menagerie," as he expressed it. Here was found the type species of the genus *At-*

185

lantosaurus (first named *Titanosaurus*). George L. Cannon, Jr., (Ref. 1) later described the large size of these bones, including a thigh bone 74 inches long and 28 inches wide at its thickest end — this bone served as a seat for three men at meals. Also found here were the genera *Apatosaurus* and *Diplodocus,* the new genus *Stegosaurus,* and *Nanosaurus* (the size of a cat), as well as the bones of crocodiles and turtles.

The horned dinosaurs were first made known to the world by discoveries in the same area. In 1887, Mr. Cannon found a pair of large, fossilized horn-cores in the Denver formation on the slopes of Green Mountain, about 2 miles east of the Morrison hogback that had come into prominence 10 years before. Whitman Cross, who was Mr. Cannon's chief on the U. S. Geological Survey party working here, sent the horns to Professor Marsh, who identified them as from ancient bison. Dr. Cross was unsatisfied with this report, because he knew the rocks to be geologically too old to contain bison fossils. But it was not until similar horns were found in Montana in 1888 by John Bell Hatcher, who was employed by Professor Marsh and soon afterward was shown more of the same kind in Wyoming, and then observed the Colorado specimens at Yale College early in 1889, that the recognition was finally made of a new group of dinosaurs having prodigious horns — a fact already suspected and commented on by Edwin Drinker Cope.

For two or three years after the first discovery, fragments of horned dinosaurs were found by Mr. Cannon, George H. Eldridge, and Dr. Cross at this and other localities in the vicinity of Denver. The same beds near Green Mountain have yielded other dinosaurs, both herbivorous and carnivorous. One, found by Mr. Cannon, was *Ornithomimus velox,* a bird-footed dinosaur about the size of a kangaroo.

Using the log and outline map published by G. Edward Lewis (Ref. 2), these historic localities may be visited as follows. Photographs are keyed to aerial photographs in *Colorado's Park of the Red Rocks. A Geologic Sketch,* by L. W. Le Roy (Denver, 1955).

Junction of Colo. 93, Colo. 74 (Alameda Parkway), and north entrance road to Red Rocks Park, about 2 miles north of Morrison. Walk south-southeast from this junction, uphill about 600 yards across the crest of the hogback along Colo. 74. This new "type locality" 'of the Morrison formation along the road cut (beneath the caprock of the Dakota formation) was the site of Professor Lakes' diggings in 1877. Of the 277 feet in thickness of Morrison, the dinosaur bones lie in gray sandstone 8 feet thick, 157 feet below the top of the formation, which begins with the first red shale, according to W. A. Waldschmidt and L. W. Le Roy (Ref. 3).

At 0.4 mile farther west on Colo. 74 can be seen ripple marks on the slope of the Dakota formation; in this road cut, dinosaur tracks have been found.

Junction of Colo. 74 (Alameda Parkway) and Rooney Gulch Road. Go south on Colo. 74 for 0.3 mile. On the right of the road for 0.2 mile are fossil deposits. First are fish fossils in the Benton formation, above the Dakota formation. Next are dinosaur tracks in the Dakota, extending 0.1 mile. Returning to the junction and continuing on Colo. 74 for 0.8 mile is the location where mammoth tusks were taken from the upper gravel when the road cut was made.

Intersection of Colo. 74 (Alameda Parkway) and Kipling Street near south entrance to Denver Federal Center. Go west on Colo. 74 for 3.8 miles; parking area on left; or it can be reached 1.9 miles from the junction of Colo. 74 and Rooney Gulch Road, described above. The horned dinosaurs were first found 0.1 mile east (toward Denver) in the ravine on the south side of the road.

Applewood Valley Methodist Church on West 20th Avenue, 0.9 mile west of Youngfield Street, at South Table Mountain, in Golden. Walk about 250 yards west to the saddle connecting the small spur and South Table Mountain. In the lower 21 feet of rock (the Denver formation) above the ravine is greenish, andesitic sandstone, in which are dinosaur and turtle bones. In the 25 feet of light, sandy clay starting 50 feet above this stratum are bones of mammals, turtles, and crocodiles. This locality has great significance in the history of geology, marking the site where Roland W. Brown solved the long-standing "Laramie problem" as to the Mesozoic-Cenozoic boundary.

Dinosaur bones, horn cores, and teeth of considerable interest have been found even within the city of Denver. As pointed out by George L. Cannon, Jr. (Ref. 4), these are isolated objects rather than entire skeletons as recovered elsewhere, because they were carried by streams rather than deposited whole in lakes.

Occasional crocodile teeth and scales have also come from the same beds.

The splendid dinosaur fossils near Canon City were discovered in 1876 by Charles E. Felch and his brother, M. P. Felch, after whom Felch Creek was named; this is the site of the collecting locality for geodes described in Segment H of *Colorado Gem Trails and Mineral Guide,* by Richard M. Pearl (Sage Books, Denver, 2d edition, 1964). Believed at first to be a section of a petrified tree, the first gigantic thigh bone was shown as a curiosity at the newspaper office in Canon City, where it was seen by the state geologist of Kansas, Benjamin F. Mudge. Thus it was, through Canon City and Denver newspapers, that the find was brought to the attention of Professor Marsh, who came to investigate.

Professor Mudge supervised the operations for a year, after which Mr. Felch was put in charge until 1886, when this historic Quarry No. 1 — known locally as the Bone Yard — was abandoned. Samuel W. Williston and Arthur Lakes made important studies here in 1877, and Mr. Felch did from 1883 to 1888. In 1901 and 1902, the quarry was reopened by W. H. Utterback and worked by the Carnegie Museum under the direction of Charles B. Hatcher. The fossils collected before 1882 are in the Yale Peabody Museum; the later ones are in the U. S. National Museum and Carnegie Museum. The U. S. National Museum acquired its specimens largely through the U. S. Geological Survey, which financed Professor Marsh's field parties near Canon City and at Como Bluff, Wyoming. His material arrived at the museum in 1898 and 1899. In the Hall of Dinosaurs and Other Fossil Reptiles is a mounted skeleton of the carnivorous dinosaur *Ceratosaurus nasicornis* Marsh from Garden Park.

No present-day collector has done so much at Garden Park as Frank C. Kessler, long-time executive secretary of the Canon City Geology Club, whose encouragement of his high-school students in their collecting and scientific activities is an inspiring story. The *Stegosaurus* from Garden Park, now mounted in the Denver Museum of Natural History, was uncovered by the

Canon City High School and excavated by Robert L. Landberg. The Municipal Museum at Sixth and River Streets, in Canon City, contains worthwhile dinosaur exhibits. The Canon City Geology Club was instrumental in the erection by the State Historical Society of Colorado of the Garden Park Dinosaur Monument, a bronze table mounted on a base of travertine; it depicts some of the reconstructions from here that are on display in various museums. The site of the marker is 8.2 miles from the museum at Sixth and River Streets, going east via River Street and north on 15th Street (toward Red Canons Park).

The bone-bearing layer of hard sandstone in this quarry is only about 3 feet thick. Yet, it yielded, in addition to interesting mammals, fish, turtles, and crocodiles, the following genera of dinosaurs: *Ceratosaurus, Labrosaurus, Stegosaurus, Morosaurus, Haplocanthosaurus, Diplodocus, Brontosaurus, Allosaurus* (properly *Antrodemus*), *Coelurus, Laosaurus,* and *Camptosaurus.*

Down Oil Creek, 0.2 miles from the monument commemorating Quarry No. 1, is the site — on the opposite side of the stream — of the recent diggings where the Cleveland Museum of Natural History removed the skeleton of *Haplocanthosaurus,* a large sauropod dinosaur.

Although Professor Marsh has been given most of the attention in this historical introduction, his brilliant and versatile rival, Edward Drinker Cope, of the Academy of Natural Sciences of Philadelphia, was also engaged at Canon City, as well as at Morrison. The bitter battle between these men, the notorious Cope-Marsh feud, was a fantastic scandal of extravagant proportions, falling barely short of open warfare. Professor Cope's collection was purchased by the American Museum of Natural History, which exhibits his *Antrodemus (Allosaurus)* from Garden Park.

A series of natural casts of dinosaur feet was taken from a coal mine near Canon City. After the coal was removed, the footprints could be seen in the overlying rock. These are on exhibit in the Denver Museum of Natural History.

SOUTHEASTERN COLORADO DINOSAUR FOOTPRINTS

The Denver Museum of Natural History has on exhibit two series of dinosaur footprints from southeastern Colorado. Those

189

of a small dinosaur, taken from the Morrison formation on the banks of the Purgatoire River, were collected by Alex Richards in 1937 near Higbee, in Otero County. Others, made by a small carnivore, came from the badlands of the same river 30 miles south of La Junta.

MIDDLE PARK DINOSAURS

Known locally as "petrified horse hoofs," broken tail vertebra of the carnivorous dinosaur *Antrodemus (Allosaurus)* occurring in Middle Park were collected by Ferdinand V. Hayden in 1869 and described in 1870 by Joseph Leidy (Ref. 5).

DINOSAUR NATIONAL MONUMENT

No place else in the world offers so spectacular an occurrence of fossil bones as have been found in Dinosaur National Monument. Most of this public reserve lies in northwestern Colorado, extending into Utah, where the Dinosaur Quarry and Visitors Center are located. From the Dinosaur Quarry and its famed Dinosaur Ledge have been taken the partial skeletons of more than 20 individuals and the scattered bones of about 300 other of the great animals.

Pieces of these bones have been picked up at Indian camp sites, and pioneer settlers and migrants in the region also knew of their existence at least as early as 1882. They were recognized as dinosaur remains in 1893 by Olaf A. Peterson, of the American Museum of Natural History. In 1908, Earl Douglass, of the Carnegie Museum, came here; the following year, on August 17, he discovered the first skeleton.

Subsequently, about 700,000 pounds of petrified dinosaur bones were taken out by the Carnegie Museum, the Smithsonian Institution, and the University of Utah. Eighty acres of the area became a national monument in 1915, and the limits were vastly enlarged in later years. In 1957, the present, unique museum — where you can watch the giant bones being carefully brought

190

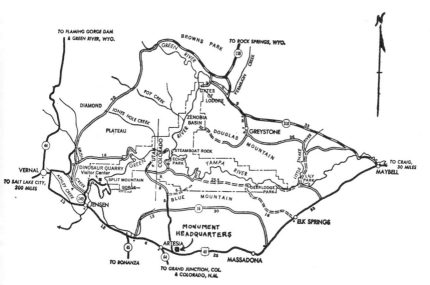

Part of Dinosaur National Monument. *National Park Service.*

to light by skilled paleontologists before your eyes — was opened to the public.

The strata of the Morrison formation stand at an angle of 67 degrees. In them are embedded the bones of *Apatosaurus,* the enormous, plant-eating dinosaur usually referred to as *Brontosaurus,* which was the first to be uncovered and the first to be mounted; it stands today in Pittsburgh, 71.5 feet long. Its companions in death — swept down an ancient stream and trapped in delta sand — included the more slender but even longer *Diplodocus,* which here reached 75.5 feet in length. The most abundant of the dinosaurs here was *Stegosaurus,* the armor-plated form that had "two sets of brains." Other dinosaurs found here are *Camarasaurus,* also a four-legged vegetarian; the tiny *Laosaurus,* only 2.5 feet long; three two-legged, plant-eating forms called *Dryosaurus, Camptosaurus,* and *Barosaurus;* and a two-footed, carnivorous kind known as *Antrodemus* (also *Allosaurus*), which here was about the size of a horse.

Mineral collectors will find agate, agatized wood, opalized wood, jasper, and agatized dinosaur bone in the region adjacent to Dinosaur National Monument, as described in Segment A of *Colorado Gem Trails and Mineral Guide.*

191

COLORADO
NATIONAL MONUMENT
COLORADO

The National Park Service publishes a free leaflet and an excellent, inexpensive booklet on Dinosaur National Monument.

COLORADO NATIONAL MONUMENT

Colorado National Monument and the surrounding region has been a source of many fine dinosaur fossils, chiefly from the upper part of the Morrison formation, known as the Brushy Basin member. In 1900, Elmer S. Riggs discoverd the first skeleton of *Brachiosaurus* at a site now marked by a bronze tablet set in a stone monument above the road cut. The place is 1.2 miles south of Colo. 340, 4.0 miles west of the intersection of Colo. 340 and U. S. 6-24-50. This dinosaur is mounted in the Chicago Natural History Museum. Now also seen in the same museum is the rear half (!) of the *Apatosaurus (Brontosaurus)* found in 1900 by Mr. Riggs; the rest of the mount is from Utah and Wyoming. The site is on Colo. 340, 1.2 miles south of the intersection of Colo. 340 and U. S. 6-24-50 (at Fruita), on the southeast side of the hill ahead of this stop. A humerus of *Brachiosaurus*, about 6 feet long, from Montrose County, is exhibited in the Hall of Dinosaurs and Other Fossil Reptiles of the U. S. National Museum, Washington.

Remains of *Antrodemus (Allosaurus), Diplodocus,* and *Stegosaurus* have also come from this general area. On the ceiling of the Thomas coal mine, near Grand Junction, can be seen, outlined for visibility, the three-toed track (34 inches in diameter) of a huge carnivorous dinosaur, perhaps *Tyrannosaurus.* The Red Mountain coal mine, near Cedaredge, was said by Charles N. Gould to contain more and larger dinosaur tracks than any comparable area in the world. Left by the iguanodont, they are preserved on the ceiling as footprints measuring 40 by 38 inches.

In the American Museum of Natural History, New York, are displayed, in inverted position, gigantic iguanodont dinosaur footprints from the roof of the Charles G. Gates coal mine at Cedaredge. These are the sandy fillings of the tracks, originally made in peat or decaying vegetation. The sand later changed to sandstone and the peat to coal. The tracks measure 34 inches in width and in length.

The best places for dinosaur bone and associated gizzard stones are west and north of Colorado National Monument, where faulting has lowered the productive strata several hundred feet below their altitude in the escarpment. Although mineral collecting is not permitted in the monument, numerous specimens occur in the surrounding country. A list of them and detailed descriptions and road logs concerning Opal Hill, Glade Park, Pinon Mesa, and the Indian Hunting Ground are given in Segment B of *Colorado Gem Trails and Mineral Guide*.

The National Park Service issues a free leaflet on Colorado National Monument.

LITERATURE ON DINOSAURS

Comprehensive, thoroughly illustrated, studies of the dinosaurs named below are given in these publications, which also contain historical information on the localities as well as the fossils.

"Osteology of the Armored Dinosauria in the United States National Museum, with Special Reference to the Genus Stegosaurus," by Charley Whitney Gilmore, U .S. National Museum Bulletin 89, 1914. Historical information on the Garden Park excavations is given on pages 24-25; but see below.

"Osteology of the Carnivorous Dinosauria in the United State National Museum, with Special Reference to the Genera Antrodemus (Allosaurus) and Ceratosaurus," by Charles Whitney Gilmore, U. S. National Museum Bulletin 110, 1920. Corrected historical information on the discoveries and excavations at Garden Park appears on page 77.

"The Ceratopsia," by John B. Hatcher, U. S. Geological Survey Monograph 49, 1907.

"Report on the Vertebrate Paleontology of Colorado," by Edward D. Cope, U. S. Geological and Geographical Survey of the Territories (Hayden Survey), Annual Report 7, 1874, pages 427-533.

"Geology of the Denver Basin in Colorado," by Samuel Franklin Emmons, Whitman Cross, and George Homans Eldridge, U. S. Geological Survey Monograph 27, 1896.

"The Dinosaurs of North America," by Othniel Charles Marsh,

Sixteenth Annual Report of the U. S. Geological Survey, 1896, p. 133-414. Summary of all American dinosaurs known at that time.

More historical information on Colorado dinosaurs, including maps, appears in these recent books of general appeal: *Marsh's Dinosaurs*, by John H. Ostrom and John S. McIntosh, Yale University Press, New Haven, 1966; *Men and Dinosaurs*, by Edwin H. Colbert, E. P. Dutton and Company, Inc., New York, 1968; and *The Day of the Dinosaur*, by L. Sprague de Camp and Catherine Crook de Camp, Doubleday and Company, Inc., New York, 1968.

The following books deal with dinosaurs only, on a popular level:

Dinosaurs, by Nicholas Hotton III, Pyramid Publications, New York, 1963. A good, very low-priced book.

Dinosaurs; Their Discovery and Their World, by Edwin H. Colbert, E. P. Dutton & Co., Inc., New York, 1961. A very interesting book.

The World of the Dinosaurs, by D. H. Dunkle, Smithsonian Institution, Washington, 1957. A simpler, good book.

URANIUM IN FOSSIL BONE

Because uranium has a chemical affinity for the carbon in organic matter, much of the Colorado fossil dinosaur bone has been impregnated or replaced by uranium minerals. Both dinosaur bones and uranium deposits are concentrated in the Morrison formation. Here, many of the deposits of petrified dinosaur remains became the sites of small uranium mines during the uranium boom of the early 1950's. The same association has resulted in the mining of a good deal of petrified wood, likewise rich in uranium minerals by infiltration and substitution (see page 159).

MARINE REPTILES

Related to the dinosaurs were Mesozoic reptiles that swam, such as icthyosaurs, plesiosaurs, and mosasaurs, as well as some that flew, such as pterosaurs.

The plesiosaurs are outstanding among the large, fishlike reptiles of the age of dinosaurs. One was described as "a snake strung through the body of a turtle." A superb example from Baca County is on exhibit in the Denver Museum of Natural History.

Representing other lizards that became adapted to life at sea were the mosasaurs, 20 feet or more of slender body and tail. The neck was short, and the elongated head was supplied with a full complement of teeth. Colorado rocks have yielded splendid specimens of some of these "sea serpents," which abandoned the hard-won land life of their ancestors and went back to the water. One of the best and largest was uncovered on the Mesa of Colorado Springs by Charles W. Warren, a student of the author's.

Marine crocodiles also appeared in the Mesozoic Era, as did marine turtles of huge size. Colorado specimens are well represented in the monumental treatise *The Fossil Turtles of North America*, by Oliver Perry Hay (Carnegie Institution of Washington, 1908), but most of them are of Tertiary age, especially from the Oligocene strata of northeastern Colorado, in which the giant land tortoise *Testudo* is abundant. The Denver Museum of Natural History has a good display from Yuma County.

REFERENCES

Ref. 1. Colorado Scientific Society Proceedings, vol. 4, 1895, p. 233-270.
Ref. 2. Guide to the Geology of Colorado, Geological Society of America *et al*, Denver, 1960, p. 285-292.
Ref. 3. Geological Society of America Bulletin, vol. 55, 1944, p. 1097-1113.
Ref. 4. Colorado Scientific Society Proceedings, vol. 8, 1906, p. 194-198.
Ref. 5. Academy of Natural Sciences of Philadelphia Proceedings, 1870, p. 3-4.

Chapter 26

Fossil Birds

Bird fossils are rare in Colorado, as elsewhere, because they are so fragile. The oldest was the one whose webbed feet made the 70 or so tracks in the sandstone of the Dakota formation of Cretaceous age, near Golden, as reported by M. G. Mehl (Ref. 1). The largest may have belonged to the genus *Diatryma,* but only plumage has been found, by Mrs. T. D. A. Cockerell in shale of the Green River formation of Eocene age.

Large vultures, a cormorant, a rail, and a plover were recovered in 1923 and 1926 from the Chadron formation of Oligocene age at Pawnee Buttes, in Weld County, by the Colorado Museum of Natural History. This is the most important bird deposit of its age in North America, as described by Alexander Wetmore (Ref. 2-3). Bones and feathers, some perfectly preserved, have come from the Florissant deposits.

Writing in 1931, Harold E. Koerner reviewed the known occurrences of fossil birds and mammals in Colorado, giving a bibliography of 42 entries (Ref. 4).

An Oligocene hawk was collected in the White River formation of northeastern Colorado by a field party of the Museum of Paleontology of the University of California in 1937, as reported by Alden H. Miller and Charles G. Sibley (Ref. 5). The most ancient record of any American quail is that of the Oligocene specimen obtained in 1948 by Edwin C. Galbreath in the White River formation of Logan County, Colorado, reported by Harrison B. Tordoff (Ref. 6).

An almost perfect fossil egg was found near Willard by W. C. Toepelman (Ref. 7).

REFERENCES

Ref. 1. Geological Society of America Bulletin 42, 1931, p. 331.
Ref. 2. Proceedings of the Colorado Museum of Natural History, vol. 7, 1927, p. 3-13.
Ref. 3. Journal of the Washington Academy of Sciences, vol. 18, 1928, p. 145-158.
Ref. 4. University of Colorado Studies, vol. 18, 1931, p. 163-176.
Ref. 5. The Condor, vol. 44, 1942, p. 39.
Ref. 6. The Condor, vol. 53, 1951, p. 203-204.
Ref. 7. Geological Society of America Proceedings, 1935 (1936), p. 394.

Chapter 27

Fossil Mammals

We live in the Cenozoic Era, the Age of Mammals, of which class (Mammalia) we ourselves are members. Mammals first appeared in the Triassic Period of Cenozoic time but waited nearly 100 million years until the decline and extinction of the reptiles, from which they descended, before undergoing their unprecedented expansion and diversification from small, insect-eating creatures to the successful forms we know today.

The oldest Colorado mammal fossil is from the Morrison formation, of Jurassic age, in which insectivores of the order Pantotheria, about the size of a large mouse, left their lower jaw bones at Garden Park, near Canon City. This is the only Mesozoic mammal known from Colorado and, according to George Gaylord Simpson (Ref. 1), the only American Jurassic mammal not from Como Bluff, Wyoming. Other insectivores resident in the state in Paleocene and Oligocene times include the first mole, *Proscapols micaenus,* discovered in Tertiary rocks in America.

Multituberculates, though worldwide in distribution, are rare Eocene fossils in Colorado rocks. Marsupials, represented by the opossum, are somewhat more common. When discovered in a cave near Ignacio, a Colorado bat fossil, of Eocene age, was the world's oldest known bat.

CARNIVORES

The carnivores are flesh-eating mammals. Among the lesser known but interesting extinct animals of Colorado's past, men-

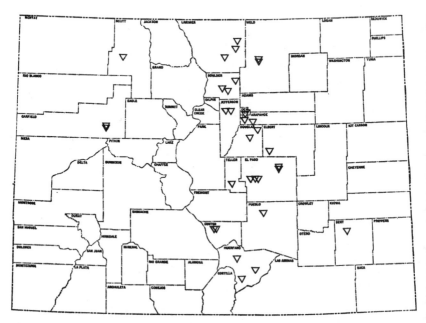

Ice-Age Vertebrate Fossil Localities in Colorado. *Oliver P. Hay, Carnegie Institution of Washington.*

tion should be made of the enormous fossil bear dog, *Amphicyon,* intermediate between a bear and a dog. Memorable examples of its bones, of Miocene age, have come from the Davis ranch, 6 miles west of Pawnee Buttes, in Weld County, as found by a Colorado Museum of Natural History expedition under Philip Reinheimer in 1920 and described by Harold J. Cook (Ref. 2).

Related to *Amphicyon,* and also found at Pawnee Buttes and elsewhere, are raccoons, bears, and other carnivores, some of doubtful identification.

Stabbing cats, *Hoplophoneus,* ancestral to the saber-toothed tiger, and *Dinictis,* predecessor of the modern cat, are known from Oligocene rocks on the Boyes ranch, near Wray, in Yuma County.

UNGULATES

The ungulates are hoofed mammals. Among the earliest of Tertiary ungulates, the condylarths were close to the ancestor of many subsequent animals of this broad group, which includes such familiar forms as horse, pig, sheep, and cattle. Colorado was included in their range, as first discovered in Moffat County by the Colorado Museum of Natural History under the field leadership of Harvey C. Markman and described by Othenio Abel and Harold J. Cook (Ref. 3). Condylarths have been taken from the Dawson formation, near Colorado Springs, together with bones of the creodont and the turtle *Conacodon,* as described by C. Lewis Gazin (Ref. 4).

Coryphodon, an amblypod, was a large ungulate familiar to Colorado scenes until it became extinct in the Eocene Epoch. It had about the size and appearance of a hippopotamus but had long, saberlike teeth.

The largest amblypod, however, was *Uintatherium,* the size of a small elephant. An exceptional find of uintathere remains in Moffat County was recovered in 1924-25 by Colorado Museum of Natural History expeditions led by Harvey C. Markman. These, described by Harold J. Cook (Ref. 5), were obtained from less than 200 feet of strata belonging to the Bridger formation in the Sand Wash Basin between Little Snake River and Vermillion Creek. The age is Eocene.

PERISSODACTYLS

Among the largest and strangest of the now-extinct mammals to reside in Colorado during the Tertiary Period were the titanotheres. These animals were perissodactyls (odd-toed, hoofed mammals), like the rhinoceros, horse, tapir, and extinct chalicothere. Appropriately named "titanic mammals," the titanotheres evolved into ponderous giants, though retaining primitive feet and teeth, and then acquired bony horns on their skulls. Not only did these developments take place almost exclusively in North America, but the great titanotheres were at time the most numerous of all mammals in the western section of the United States. Then they died out suddenly in the Oligocene Epoch. The Denver Museum of Natural History has a splendid exhibit of Oligocene skeletons from Pawnee Buttes, where the Chadron formation — so richly endowed with titanothere remains in adjoining states — outcrops. The American Museum of Natural History recovered important titanothere bones from the Huerfano Basin of south-central Colorado, as reported in 1919 by Henry Fairfield Osborn (Ref. 6).

An "astonishing variety" of titanothere remains of Eocene age was found in the Sand Wash Basin, in Moffat County, by expeditions in 1924-25 of the Colorado Museum of Natural History, led by Harvey C. Markman. The bones came from less than 200 feet of the Bridger formation, in which nine-tenths of the fossils were of titanotheres, the rest being of uintatheres and one bit of rhinoceros jaw, as described by Harold J. Cook (Ref. 7).

Work by Malcolm C. McKenna (Ref. 8-9) in the Sand Wash Basin since about 1952 has brought to light a wealth of Eocene vertebrates from the Wasatch and Knight formations. These include about 9,000 complete mammal teeth recovered by an underwater screening process.

Dr. McKenna (Ref. 10) has mapped the various quarries operated by the Museum of Paleontology of the University of California. C. Lewis Gazin (Ref. 11) has summarized the paleontologic explorations in basins adjacent to the Uinta Mountains.

Related to the titanotheres, the extinct chalicotheres (also perissodactyls) were much like massive horses having claws instead of hoofs. The claws were seemingly used for digging up roots along streams. Bones have come from northeastern Colorado.

Miocene-age tapirs are another kind of perissodactyl found fossil in Colorado, as at the Davis ranch, in Weld County.

Another perissodactyl group, the rhinoceroses, seem unlikely inhabitants of Colorado. But, during the Tertiary Period, they thrived in this state, though never reaching the 17-foot size of their Asiatic cousin *Baluchitherium,* which was, next to the dinosaurs, the largest land animal that ever lived.

The Tertiary rocks of northeastern Colorado were first investigated by O. C. Marsh, who visited Weld County in 1870 and noted the "Titanotherium beds" and, above them, the oreodont beds (Ref. 12). E. L. Berthoud, while searching for ancient man in 1871, reported fossils present in the Crow Creek area (Ref. 13). Edwin D. Cope (Ref. 14) studied the geology and collected fossils in 1873 and 1879. A Princeton University expedition collected near the Chalk Bluffs, familiar to Professor Marsh, in 1882. The American Museum of Natural History sent field parties to Weld and Logan Counties in 1898, 1901, and 1902, led by W. D. Matthew, Barnum Brown, and A. Thomson (Ref. 15-16).

The most rewarding locality in Colorado for the collecting of vertebrate fossils — and one of the finest in the world — has been Pawnee Buttes, the twin mesas that rise above the valley of the South Platte River in Weld County. Isolated by erosion from the southward-facing bluffs that extend east-west near the boundary of Wyoming and Nebraska, East and West Pawnee Buttes are composed of Tertiary sedimentary rocks: the Arikaree formation (sandstone and gravel) of Miocene age at the top, the White River formation of Oligocene age beneath. The White River is separated into two members: the Brule (mostly clay) above the Chadron. In the Chadron member have been found the remains of nearly 100 species of land vertebrate animals. From here, we have learned much about the evolution of the horse and camel. It is, however, the two *Trigonias* rhinoceros quarries, opened on Horsetail Creek in 1920, that rank as the most significant mammalian finds made in Colorado for the first three-quarters of a century.

The University of Colorado came briefly in 1907 (Ref. 17). The chief results of this early work was the recovery of excellent bones of Tertiary horses at the Davis ranch.

In an area passed up by other museums, an expedition of the Colorado Museum of Natural History was camped in 1920. It consisted of Philip Reinheimer, Frank Howland, and Harvey C. Markman. Their search seemed futile, the surface bones, or "float," promising little because their source was unknown.

"Then on a hot Sunday afternoon Markman climbed a little barren hill near camp in search of a breeze and some relaxation. The hilltop was a network of deeply cut gullies all of which had been scanned carefully many times before. Here and there a crumbling specimen of rhinoceros bone remained in the hard, limy walls, none of them worth taking out. Presently it began to dawn upon the meditating man that things were not reasoning out as they should. On the flat, immediately below, there was surface bone of much better quality than anything weathering out of that hill, yet there was no other elevation from which the supposed 'float' could have been carried. There could be only one good explanation: the softer formation at the base of the hill was undergoing slow erosion at the surface, mainly by the wind, and the loose scrap had not been transported at all. Any remaining source of supply had to be underneath and at no great distance.

"Results of the first digging were not too encouraging and it began to look as if the argument would not sustain itself. However, a few scattered bones were eventually uncovered and in every case they were in excellent condition. Perseverance was rewarded sometime later when the first string of articulated vertebrae was uncovered. Next a skull was found and from then on the going was fine.

"This quarry was operated for a number of years with the most gratifying results. On the side of luck may be mentioned the fact that when Mr. Figgins said the quarry ought to produce some titanothere material, in addition to the abundance of rhino skeletons, it was only a short time until titanotheres were on the assembly line; and when the request came for entelodonts, there was little delay — entelodonts were produced forthwith." — Annual Report, 1942, Colorado Museum of Natural History.

The circumstances of the discovery told by T. D. A. Cockerell (Ref. 18) are in error, the above account being more accurate. This bone deposit was once a muddy waterhole, or slough. The rhinoceros *Trigonias*, of which 35 skulls were obtained, was a rather primitive genus having a reduced canine tooth and three incisors in the upper jaw. Descriptions were given by William K. Gregory and Harold J. Cook (Ref. 19) and by Horace Elmer Wood, 2nd (Ref. 20).

Between 1923 and 1940, collecting was done by the Colorado Museum of Natural History, the University of California, Childs Frick, and the University of Kansas. In 1940, G. Edward Lewis and Robert W. Wilson worked here; and in 1946, C. W. Hibbard led a University of Kansas field party to the region. Edwin C. Galbreath (Ref. 21) has reviewed the history of these investigations, except those by Colorado institutions; his report has a bibliography of 190 entries. The systematic paleontology is outlined by geologic epoch, and the faunas are listed for the localities named Horsetail Creek, Cedar Creek, Vista, Martin Canyon, Pawnee Creek (divided into Eubanks, Kennesaw, and Vim-Peetz), and Sand Canyon. The fossils described include fish, amphibians, reptiles, birds, and mammals of considerable diversity.

Ogallala strata of Pliocene age near Wray, in Yuma County, yielded important mammal fossils to field parties of the Colorado Museum of Natural History under the leadership of Philip Reinheimer, in 1918-1919. The site was on the ranch of H. D. Boyes, who found it. Besides the short-legged rhinoceros *Teleoceras*, the fauna included mastadons, three-toed horses, camels, large dogs, stabbing cats, peccaries, and other animals. Harold J. Cook (Ref. 22-25) has described the discoveries here. A remarkable mount of *Teleoceras* in the Denver Museum of Natural History shows the skeleton on one side and a modeled restoration showing on the other side how the creature looked when alive.

HORSES

In our study of the evolution of life, no group of mammals exceeds the horses (which are perissodactyls) in the completeness of the available fossil record. From *Eohippus*, the "dawn horse," to *Equus*, the horse of today, the steadily changing forms are preserved in the successive layers of sedimentary rocks. Here are revealed the gradual development from four toes (and undoubtedly from a five-toed predecessor) to one toe; from teeth equipped for browsing, to those adapted to grazing; and of body size, length of leg, straightness of back, and skull structure to accommodate the grazing type of teeth. It is interesting to realize that, although the horse evolved as a dominantly North American animal, it had disappeared entirely from this continent before the coming of man and was brought back by the Spanish conquerors. This unexplained mass extinction affected other large animals in North America — including mammoth, mastodon, camel, ground sloth, glyptodont, and giant armadillo — at about the same time.

Of the discoveries of early horses in Colorado, the find by J. L. Wortman of the skeleton of *Eohippus* upon which the first complete description of a four-toed horse was made ranks high in scientific significance. The date was 1880, the place was Huerfano Park (which lies in Huerfano County, north of the Spanish Peaks), and the describer was Edward Drinker Cope, of dinosaur fame. This specimen, standing 12 inches high at the shoulder, is in the American Museum of Natural History in New York. The Huerfano formation, of Eocene age, was first described in 1886 by R. C. Hills (Ref. 26) and explored in 1897 by Henry Fairfield Osborn and J. L. Wortman (Ref. 27). The fossils occur in a layer 10-15 feet thick, situated 30-40 feet above the base of the formation. Besides the horse *Eohippus*, there are also present *Coryphodon* and the remains of camels and giant turtles.

Of outstanding importance among Colorado horses were those that died at Pawnee Buttes. This remarkable fossil locality has revealed invaluable knowledge about the sequence of horse evolution. In 1901, it furnished the finest known skeleton of three-toed, Miocene *Hypohippus*, the "forest horse," which was found by Barnum Brown and placed in the American Museum of Natural History. From here, also, came a dwarf, three-toed horse, *Parahippus*, of Miocene age; the three-toed, grazing *Merychippus* (Miocene), found by Edward Drinker Cope and now in the American Museum of Natural History; and another, from the same stratum, found in 1919 and now in the Amherst College Museum.

Oliver P. Hay (Ref. 28) enumerated finds of fossil horses made in nine Colorado counties up to 1924.

For further information on fossil horses, on a popular level, see:

Evolution of the Horse, by W. D. Matthew and S. H. Chubby, American Museum of Natural History, New York, 7th edition, 1932.

Fossils. A Story of the Rocks and Their Record of Prehistoric Life, by Harvey C. Markman, Denver Museum of Natural History, 4th edition, 1961.

Horses, by George Gaylord Simpson, Oxford University Press, New York, 1951.

The Evolution of the Horse, by Frederic Brewster Loomis, Marshall Jones Company, Boston, 1926.

Like the horses, the camels evolved mainly in North America and then became extinct on this continent until reintroduced by man. Colorado camels show a substantial record of their changing characteristics during the Tertiary Period. Camels are artiodactyls, or even-toed, hoofed animals, like the hippopotamus, giraffe, deer, cattle, goat, and sheep. Fossils are known from Pawnee Buttes, Huerfano Park, Wray, and miscellaneous localities. Oliver P. Hay (Ref. 29) enumerated finds of camel in four counties in Colorado.

Enteledonts are ancient swine known wrongly as giant pigs, which lived abundantly in Tertiary Colorado, and their bones have been found associated with those of the rhinoceroses and titanotheres. Mounted exhibits in the Denver Museum of Natural History show their habitat. They were artiodactyls as large as buffalo and had knobs protruding from their yard-long skulls, both on top and underneath. Pawnee Buttes is a major locality for these beasts.

Colorado artiodactyl fossils also include those of the giraffe camel *Alticamelus* (from Pawnee Buttes), as well as a musk-ox from near Colorado Springs (Ref. 30), extinct species of bison from four counties (Ref. 31), a Miocene deer, and the common oreodont. Oreodonts, among the most abundant of extinct hoofed mammals of Tertiary time, were well represented in Colorado, as at Pawnee Buttes, in Weld County.

MAMMOTHS AND MASTODONS

These interesting members of the elephant family — the Proboscidians, which are grouped as subungulates — roamed Colorado during the Ice Age, becoming extinct in North America between the times of the arrival of the Indian and the coming of the European. Elephant ivory from the tusks of mammoths and mastodons is a well-known type of fossil in the unconsolidated Pleistocene sediments of Colorado. With the cooperation of Father Conrad Bilgery, of Regis College, the remains of a dozen or more mammoths were recovered near Dent, in southwestern

Weld County, by the Colorado Museum of Natural History in 1933. One skeleton was mounted by the museum, and another was traded to the Carnegie Museum as partial payment for the great *Diplodocus* from Dinosaur National Monument.

The teeth of these animals — which is the chief means of telling them apart — are likewise fairly common. Mammoth teeth are high crowned, like those of the modern elephant, and have a closely ridged pattern of flattened, enamel plates. Mastodon teeth are low crowned and show cusps; Colorado specimens are known as old as the Miocene Epoch.

Elephant teeth have been found in various excavations in and near Denver. One tooth, formerly in the Chamber of Commerce Museum, was found in 1915 at Sixteenth and Larimer Streets; another, once on exhibit in the old Denver High School, came from Seventeenth and California Streets, together with bones of a camel. Many have been recovered in more recent years.

Altitude is no barrier, for the teeth of mammoths have been found in the high mountains of southern Colorado. Harld J. Cook called attention to them in 1930 (Ref. 32) and 1931 (Ref. 33). Mammoths and bison are associated with the artifacts of early man in Yuma County; the extinct *Bison taylori* and camel are found similarly at the famous Lindenmeier site, 28 miles north of Fort Collins. The Lamb Spring site at Littleton, uncovered in 1963, revealed artifacts in association with the remains of the horse, bison, camel, and Columbian mammoth (Ref. 34).

Oliver P. Hay (Ref. 35) enumerated finds of the Columbian elephant *(Elephas columbi)* made in nine counties up to 1924, of the Imperial elephant *(Elephas imperator)* in six counties, and of undetermined species of elephants in five counties.

GROUND SLOTHS

Few extinct animals were stranger than the ground sloths: huge, plant-eating mammals equipped with teeth on the sides of their jaws, and large claws for digging up roots and tearing down branches. They walked clumsily on the outer edge of their feet. The ground sloth was a member of the order of mammals known as the edentates, or "toothless mammals," a description that fully

208

applied to certain of them, although the rest had a very simple tooth structure. Evolving from insectivore ancestry, the edentates gave rise to the ground sloth, as well as to the extinct glyptodonts and the still-living tree sloth, anteater, and armadillo. The ground sloth developed in South America, invading the northern continent during the Ice Age.

The discovery in Pleistocene sediments of a fossil ground sloth near Walsenburg by T. D. A. Cockerell (Ref. 36) was a noteworthy one. The skull measures 22 inches in length. It was found by E. A. Lidle on his farm 1 mile south of Walsenburg and given to the University of Colorado Museum by Mrs. Lidle. Comparison with other ground sloth remains suggest that the animal is the genus *Mylodon,* as large as a rhinoceros.

A second Colorado ground sloth, a femur but not petrified, was found along Bijou Creek 5 miles north of Agate by Dale Purdy and was given to the Denver Museum of Natural History in 1965. A third one has been found near Security.

RODENTS

Remarkably advanced among mammals, rodents are gnawing animals. Among fossil rodents of Colorado is a horned gopher, *Ceratogaulus rhinoceros,* found by Barnum Brown on an expedition in 1901 of the American Museum of Natural History at Loup Fork, in Weld County, in the Pawnee Creek formation (now the Arikaree) of Miocene age. W. D. Matthew described it as having "a profile absurdly like that of a miniature rhinoceros," and so gave it the name.

MAPS AND LITERATURE ON MAMMALS

In 1924, Oliver P. Hay summarized (Ref. 37) the discoveries of Pleistocene vertebrate fossils in western North America, describing those of each kind and each state separately. A series of maps shows the localities for each kind of animal.

In 1910, Henry Fairfield Osborn listed Glenwood Springs as

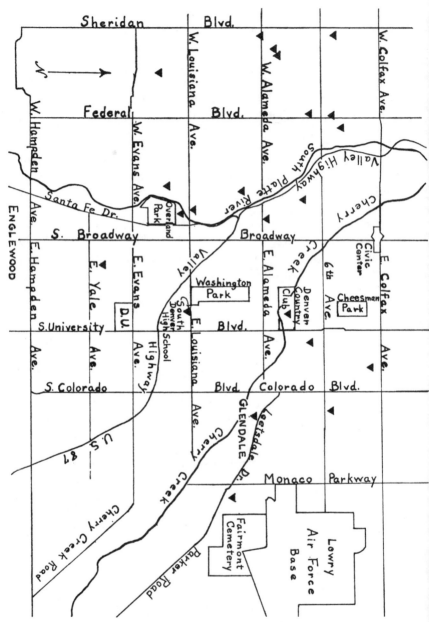

Ice-Age Fossil Localities in Denver Area, South of Colfax Avenue.

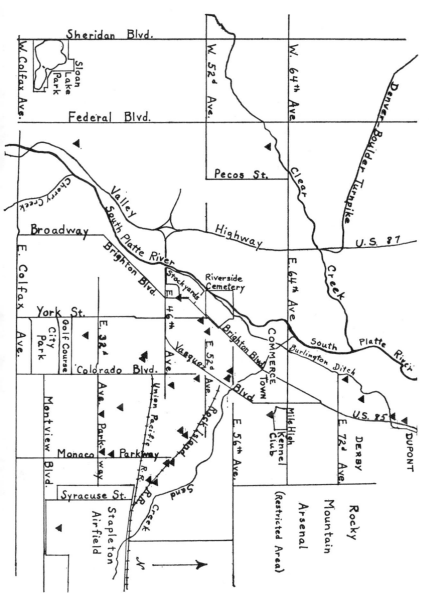

North of Colfax Avenue. *Charles B. Hunt, U. S. Geological Survey.*

one of the 33 chief Pleistocene mammal localities in North America. South Canyon, 5 miles west of the city, is mentioned a few times in the literature, but later discoveries elsewhere have greatly overshadowed it.

On Plate 3 of the U. S. Geological Survey Bulletin 996-C, Charles B. Hunt has mapped the places in the Denver area where more than 132 collections of Pleistocene fossil mammals have been made. These include 19 species of antelope, deer, elk, bison, camel, horse, mammoth, musk-ox, prairie dog, ground squirrel, wolf, and fox. Only two of the places represented fairly complete skeletal resources, the rest being composed of fragmentary material. A Pleistocene horse, camel, and bison found 0.5 miles south of Fairmount Cemetery, in Denver, are preserved in the U. S. National Museum, Washington, as described by Oliver P. Hay (Ref. 38). Glenn R. Scott has mapped 60 fossil-vertebrate localities in the Kassler quadrangle, Douglas and Jefferson Counties (Ref. 39).

The summary of fossil mammals and birds in Colorado written in 1931 by Harold E. Koerner (Ref. 40) includes a bibliography of 42 references.

The complete, detailed bibliographies on vertebrates, indexed by classification but not by occurrence, are the following publications:

U. S. Geological Survey Bulletin 179, 1902.
Carnegie Institution of Washington Publication 390, volume 1, 1929; volume 2, 1930.
Geological Society of America Special Paper 27, 1940.
Geological Society of America Special Paper 42, 1942.
Geological Society of America Memoir 37, 1949.
Geological Society of America Memoir 57, 1953.
Geological Society of America Memoir 84, 1961.
Geological Society of America Memoir 87, 1963.
Geological Society of America Memoir 92, 1964.

In addition to textbooks on historical geology, vertebrate fossils are treated in the following books:
Evolution of the Vertebrates, by Edwin H. Colbert, John Wiley and Sons, Inc., New York, 1955. A college textbook.
Search for the Past, An Introduction to Paleontology, by

James R. Beerbower, Prentice-Hall, Inc., Englewood Cliffs, N. J., 2d edition, 1968. Covers both vertebrate and invertebrate fossils.

The Vertebrate Story, by Alfred Sherwood Romer, University of Chicago Press, 1959. An interesting book on fossil and modern vertebrates.

Vertebrate Paleontology, by Alfred Sherwood Romer, University of Chicago Press, 3d edition, 1966. An illustrated, college textbook.

REFERENCES

Ref. 1. Memoirs of the Peabody Museum of Yale University, vol. 3, pt. 1, 1929.

Ref. 2. Proceedings of the Colorado Museum of Natural History, vol. 6, 1926, p. 29-31 and 1 plate.

Ref. 3. Proceedings of the Colorado Museum of Natural History, vol. 5, 1925, p. 33-36.

Ref. 4. Journal of the Washington Academy of Sciences, vol. 31, 1941, p. 289-295.
Rocky Mountain Association of Geologists, 14th Field Conference Guidebook, 1963, p. 167-169.

Ref. 5. Proceedings of the Colorado Museum of Natural History, vol. 6, 1926, p. 7-11 and 5 plates.

Ref. 6. American Museum of Natural History Bulletin, vol. 41, 1919, p. 557-569.

Ref. 7. Proceedings of the Colorado Museum of Natural History, vol. 6, 1926, p. 12-18 and 11 plates.

Ref. 8. Geological Society of America Bulletin, vol. 65, 1954, p. 1283.

Ref. 9. Journal of Mammalogy, vol. 35, 1954, p. 581.

Ref. 10. Intermountain Association of Petroleum Geologists, 6th Annual Field Conference Guidebook, 1955.

Ref. 11. Intermountain Association of Petroleum Geologists, 6th Annual Field Conference Guidebook, 1955, p. 41-43.

Ref. 12. American Journal of Science, ser. 2, vol. 50, 1870, p. 292.

Ref. 13. Academy of Natural Sciences of Philadelphia Proceedings, 1872, p. 46-49.

Ref. 14. U. S. Geological and Geographical Survey of the Territories Annual Report, 1873, p. 427-533 and 8 plates.

Ref. 15. American Museum of Natural History Memoirs, vol. 1, pt. 7, 1901, p. 355-447 and 3 plates.

Ref. 16. American Museum of Natural History Memoirs, new series, vol. 2, pt. 1, 1918, p. 1-331.

Ref. 17. University of Colorado Studies, vol. 4, 1907, p. 145-165.

Ref. 18. Scientific Monthly, vol. 17, 1923, p. 271-277.
Zoology of Colorado, University of Colorado, Boulder, 1927, p. 4.

Ref. 19. Proceedings of the Colorado Museum of Natural History, vol. 8, 1928, p. 2-32 and 6 plates.

Ref. 20. Journal of Mammalogy, vol. 12, 1931, p. 414-428.

Ref. 21. University of Kansas Paleontology Contributions. Vertebrata, Art. 4, 1953.

Ref. 22. Proceedings of the Colorado Museum of Natural History, vol. 4, 1922, p. 2-14.

Ref. 23. Proceedings of the Colorado Museum of Natural History, vol. 4, 1922, p. 3-29.

Ref. 24. Proceedings of the Colorado Museum of Natural History, vol. 7, 1927, p. 1-5.

Ref. 25. Proceedings of the Colorado Museum of Natural History, vol. 9, 1930, p. 44-51 and 7 plates.

Ref. 26. Proceedings of the Colorado Scientific Society, vol. 3, 1889 (1890), p. 148-164, 217-223.

Ref. 27. American Museum of Natural History Bulletin, vol. 9, 1897, p. 247-258.

Ref. 28. Carnegie Institution of Washington Publication No. 322A, 1927, p. 143-145.

Ref. 29. Carnegie Institution of Washington Publication No. 322A, 1927, p. 166-167, 355.

Ref. 30. Carnegie Institution of Washington Publication No. 322A, 1927, p. 182, 361.

Ref. 31. Carnegie Institution of Washington Publication No. 322A, 1927, p. 195-196, 363.

Ref. 32. Science, vol. 72, 1930, p. 68.

Ref. 33. Science, vol. 73, 1931, p. 283-284.

Ref. 34. American Antiquity, vol. 29, 1963, p. 14.

Ref. 35. Carnegie Institution of Washington Publication No. 322A, 1927, p. 72-74, 96-97, 111-112.

Ref. 36.   University of Colorado Studies, vol. 6, 1909, p. 309-312 and 2 plates.
Ref. 37.   Carnegie Institution of Washington, Publication No. 322A, 322B, 1927.
Ref. 38.   Proceedings of the U. S. National Museum, vol. 59, 1921, p. 599-603.
Ref. 39.   U. S. Geological Survey Professional Paper 421-A, 1963.
Ref. 40   University of Colorado Studies, vol. 18, 1931, p. 163-176.